The Best
AMERICAN
SCIENCE &
NATURE
WRITING
2023

GUEST EDITORS OF THE BEST AMERICAN SCIENCE AND
NATURE WRITING

2000 DAVID QUAMMEN
2001 EDWARD O. WILSON
2002 NATALIE ANGIER
2003 RICHARD DAWKINS
2004 STEVEN PINKER
2005 JONATHAN WEINER
2006 BRIAN GREENE
2007 RICHARD PRESTON
2008 JEROME GROOPMAN
2009 ELIZABETH KOLBERT
2010 FREEMAN DYSON
2011 MARY ROACH
2012 DAN ARIELY
2013 SIDDHARTHA MUKHERJEE
2014 DEBORAH BLUM
2015 REBECCA SKLOOT
2016 AMY STEWART
2017 HOPE JAHREN
2018 SAM KEAN
2019 SY MONTGOMERY
2020 MICHIO KAKU
2021 ED YONG
2022 AYANA ELIZABETH JOHNSON
2023 CARL ZIMMER

The Best AMERICAN SCIENCE & NATURE WRITING™ 2023

Edited and with an Introduction
by Carl Zimmer

Jaime Green, *Series Editor*

MARINER BOOKS
New York   Boston

FIRST EDITION

ISSN 1530-1508

ISBN 978-0-06-329321-2

23 24 25 26 27 LBC 5 4 3 2 1

Contents

Foreword

THIS IS MY fifth year as series editor of this anthology. I started in 2019, which now only sounds to me like "the year before 2020." So that was the first year; the year before. The 2020 anthology collected work published in 2019; I wrote my foreword in the first months of the pandemic, about a book full of The Before; that was the second year. The 2021 anthology, fittingly guest-edited by Ed Yong, collected writing from the first year of the pandemic: the shock, the race for knowledge, the harrowing human effects. So my third year was the year it hit. And last year was the second pandemic year. This year, my fifth, and in writing the pandemic's third, is a curious one to try to label. Maybe like labeling 2019 The Year Before, it'll be possible only in hindsight. (Hopefully not as The Year Before anything else.) But as this year's guest editor Carl Zimmer writes in his introduction, it's a year that's trying to transition, where the pull out of pandemic panic into something longer term is happening in society and in policy. The pandemic isn't over, but the official emergency is; so as people and as writers and readers we're working to figure out what that means.

One grim way of looking at it is in terms of how the pandemic subsides to make room for other horrors. Even in 2020, of course, there was no turning away from the murder of George Floyd and the outcry and movement that followed. (It's pure privilege to even consider that an event that could be turned away from, rather than the manifestation of forces always at play in your daily life.) And climate change stops for no one, unless one does the work to stop *it*. This year, two of the most important American stories—the repeal of *Roe v. Wade* and the assaults on trans rights—hinge on science, too. They hinge, though, on its perversion, medical knowledge and terminology wielded in service of tyranny and fascism.

(I think fondly of those early years when I'd get a handful of emails complaining that the series had gotten too political. It's what I wrote my foreword about once, before there was a pandemic to worry about. Those emails have quieted down lately, maybe because the political nature of science is impossible to ignore now, or maybe those complainers have given up on reading. We'll see who comes out of the woodwork for me calling anti-abortion and anti-trans political projects fascist. But they're about controlling marginalized bodies and dictating what kinds of lives are acceptable. Feel free to email me, I won't be responding if it's about that!)

A less grim lens through which to see the shift to the pandemic's sharing space with other stories is the power of science—the utility and vitalness of work being done to understand the world, and the power and beauty of writing about that work, of telling its stories, highlighting its practitioners, revealing the ways it weaves through our daily lives. Science helps us understand the world, appreciate its beauty, deepen our connections to nonhuman life, etc. etc. etc.

That's the only kind of science writing those it's-too-political complainers want us to focus on. But it's in these pages alongside explicitly political work, because the best science and nature writing comprises *all* of that.

With political questions in mind, I turn to the power of first-person writing. This is my bias as an essayist coming through. But in first-person writing, science writers are not only reporters and translators but narrators, too. Narrators of their own experience, of their own pain and love and questions. Where amid showing us the knowledge of science and the beauty of nature they also say, *This affects human lives, including mine. Here's how.*

In their essay "My Metamorphosis," Sabrina Imbler writes of the transcendent power of transition, drawing resonance with and inspiration from the myriad ways insects metamorphose. "I deserve to be comfortable," they write. "I deserve to feel at home in my body and what surrounds it. Because in spite of all the voice cracks and pimples and other embarrassments of a second puberty, I have never felt lighter, more free." It is as natural as a caterpillar spinning a cocoon, as violent as the hour a mayfly larva stops breathing while it molts, as personal and as beautiful as filling your lungs with fresh air.

Where Imbler softly celebrates the ownership of one's own body, Annie Lowrey's "American Motherhood" is an eloquent animal cry of pain at what it feels like to lose the very same thing. Lowrey describes in vivid detail two harrowing, mysteriously challenging pregnancies, beset by maddening itching, preeclampsia, and traumatic deliveries, to which she points and says, *This is what the Supreme Court thinks you should be forced to endure.* Lowrey quotes a physician who asks, "What actually counts as the life of the mother? Is that her life today? Is that her life during this pregnancy?" The right to abortion isn't just about a pregnant parent not dying, of course, it's about not being forced to be pregnant. Lowrey vividly and viciously evokes the multifarity of pregnancy, the love and hope, the fear, the physical agony. And the lostness in the wake of *Dobbs*: "What does the life and health of mothers mean? How could it possibly mean so little? What are we supposed to do?"

I don't have a good answer. In their essay, Imbler writes, "the more famous kind of metamorphosis is more total—that quintessential transmogrification from a caterpillar to chrysalis to butterfly." I hope that such a transformation may be in store for us all, as individuals to live the lives we choose, and for our society to shed its constrictive shell, toward a new way of being. "It is a promise to trust the process, that even mired in goo and guts, something like bliss may wait on the other side." I hope we are on our way there.

Submissions for next year's edition of The Best American Science and Nature Writing are open. Visit jaimegreen.net/basn for information and to nominate work, including your own.

JAIME GREEN

Introduction

I'M WRITING THIS in May 2023, a month marked by two nearly identical announcements. The World Health Organization kicked things off by declaring that COVID-19 is no longer a "public health emergency of international concern." The United States followed up a few days later, announcing that the country's Public Health Emergency is over.

Is the pandemic finished? No. Three years and four months after it emerged in Wuhan, SARS-CoV-2 is still killing a few thousand people each week. The ranks of survivors with Long COVID—by one estimate, sixty-five million people—continue to grow, while scientists scramble to find treatments for their suffering. The coronavirus continues to evolve, leaving open the possibility that a new variant will emerge that evades the world's immunity and wreaks fresh havoc. Instead, what is ending is the way we treat the pandemic. The world is going to begin treating it as yet another feature on the medical landscape, no different than cancer, heart attacks, and seasonal influenza.

While the pandemic may not actually be over, this shift in the public's attention is a significant moment for science writers. For the past three years and change, our community focused much of our effort on a single virus. Now, as we widen our gaze again, it's a good time to consider what we learned about the craft during the pandemic and what lessons we can apply to the stories that lie ahead.

Looking back, it's hard to believe how uncertain COVID-19

looked at the very beginning. But in January 2020, we could not yet know for sure what we were dealing with. Two other coronaviruses had emerged over the previous two decades, and if SARS and MERS were any guide, the new one would be deadly but containable. Late in January, I met some fellow science writers at a midtown bar—in those final few weeks before all the bars closed down—and talked about whether the new coronavirus would follow the path of its predecessors. As I sipped a stout, I heard the word I did not want to hear: *pandemic.* Apparently, the new coronavirus, SARS-CoV-2, was behaving differently. It killed a smaller fraction of its victims. But it spread far more easily.

By February, the coronavirus was running rampant in Europe and starting to flare up in the United States. My editor at the *New York Times* emailed me to let me know he was only editing stories about COVID-19. If I had ideas for other pieces—about Neanderthals, about exoplanets, about anything other than the pandemic—I would need to pitch someone else. Instead, I joined the pandemic team.

I imagined spending a few weeks pitching in. I had no idea that I was entering a frenzied state that would stretch for more than two years, during which I wrote about SARS-CoV-2 and little else. It was bizarre to find that even my most esoteric interests in biology were in high demand. When vaccines were put into trials, I described how messenger RNA works. When Omicron arrived with dozens of mutations, I explained the concept of epistasis.

The pandemic was a story far bigger than epistasis and messenger RNA, of course. Many of my colleagues tackled the other dimensions. The divergent toll of COVID-19 in its first year only made sense by taking into account the structural racism of the United States and the country's broken health care system. After a vaccine became available, a new divergence emerged, between communities that shunned vaccines and those who accepted them. To explain that shift, journalists needed to consult social scientists, not virologists.

Month after month, the story of the pandemic refused to end, no matter how much people wanted it to be over. On March 16, 2020, President Trump announced a fifteen-day plan to pull the country out of the COVID-19 crisis. "I'd love to have it open by Easter, okay?" he said. When that failed, the country's hope turned to vaccines. That hope that seemed to be borne out a few months later when clinical trials showed that they performed with impressively high efficacy. President Biden made promises of his own. By the summer of 2021, he said, Americans might reach a place "where we begin to declare our independence on Independence Day from the virus." In his fight against the pandemic, Biden relied too heavily on vaccines alone. He did not reckon with things like evolution. New variants emerged and helped to dash his summertime hopes.

The most peculiar desire for an end to the pandemic came in late 2021, when Omicron arrived. In South Africa, which saw the first surge of cases, a relatively small fraction of people ended up in the hospital. That left people with the impression Omicron was "mild." This was a simple arithmetical error. Omicron spread far faster than earlier variants. A small fraction of a much bigger number of cases still led to a tremendous spike of hospitalizations.

In 2022, the pandemic beat slowed down at last. There were still stories to write—about studies that put the estimated global death toll at fifteen million and counting, about new boosters and new antivirals. As Omicron spun off new subvariants, I would jump in to write about BA.5, XBB.1.5, XQ.1. But in 2022, there was time for other things.

The stories I've picked for this year's edition of the *The Best American Science and Nature Writing* are largely made up of those other things. But not entirely.

In "The Provincetown Breakthrough," Maryn McKenna looks back at the July 2021 outbreak in Massachusetts that shattered the country's dream of a "hot vax summer." The story McKenna tells is not one of helpless terror, however. She found an inspiring tale of a community that came together with public health authorities to prevent an outbreak from spreading further. If a new variant causes another outbreak in the months or years to come, we can only hope that we can follow Provincetown's example.

The pandemic was accompanied by many other disasters big and small. Not until the arrival of COVID-19 did I hear the term "moral injury." In "An Invisible Epidemic," Elizabeth Svoboda details the agony of medical professionals who could not live up to their own standards as the American health care system buckled under the pressure of COVID-19. A psychological condition first documented in wartime, it has now found many new victims, who are now struggling to recover.

One sign of the pandemic's vast scope was how it changed the atmosphere. With cars off the roads and factories sitting idle, humans emitted far less carbon dioxide into the atmosphere. That pause was short-lived, however. By the end of 2020, carbon emissions had rebounded, and by 2022 we set a new record. Global warming was a crisis before the pandemic began, and it remains so now.

I picked half a dozen stories of climate change for this anthology. But they are markedly different from one another, as befits a story that encompasses the planet. In "The Coming Collapse," Douglas Fox reports from an Antarctic ice shelf melting before our eyes. In "An Ark for Amphibians," Isobel Whitcomb recounts the search for refuges where species endangered by climate change may escape extinction. In "Don't Look Down," Lois Parshley reports from Alaska, where permafrost is disappearing and taking a way of life with it. In "Dislodged," Josh McColough recounts a road trip with his daughter on which climate change, COVID-19, and other burdens of this age made themselves impossible to ignore.

COVID-19 did not just lower carbon emissions. It also weakened the world's efforts to fight other diseases that grind on steadily, year in and year out. In one study, researchers found that diagnoses of cancer dropped by 13 to 23 percent in the United States—a dangerous slip that could dim many people's chances of survival in the years to come. Tuberculosis had been declining for years, thanks to global efforts to prevent its spread. In 2021, the rate jumped 4.5 percent, with an estimated 10.6 million people falling ill.

These diseases are important subjects for stories, even if it's hard to find a fresh news hook that will grab people's attention. It can be tempting to resort to hype, to drum up false hope. More than fifty-five million people have dementia, a figure that will keep climbing in years to come. That does not mean that writers should dangle the prospect of potential cures—especially when the most promising drugs for Alzheimer's disease show only modest benefits and potentially lethal side effects. In "A French Village's Radical Vision of a Good Life with Alzheimer's," Marion Renault discovers a different story to tell about dementia. She reports on how to organize the lives of people with Alzheimer's to bring them comfort.

While Alzheimer's disease is all too familiar, other medical conditions are only now coming to light. Fletcher Reveley investigates a new ailment that may turn out to be a major burden in Latin America in his story "In El Salvador and Beyond, an Unsolved Kidney Disease Mystery." We must also be on our guard about labeling certain aspects of our lives as medical disorders. Sarah Zhang, in her piece "The Myopia Generation," looks at the rising tide of shortsightedness and questions whether medicalizing it with eye drops and other treatments is the right way to respond to this change in our visual lives.

The pandemic also demonstrated that we cannot understand things in the world without understanding how we behave in relation to them. A COVID vaccine is a scientific object, of course, but it came into existence only thanks to a long string of decisions made by individuals, companies, and governments. COVID vaccines were developed in record time not because they were easy, but because—among other things—the National Institutes of Health decided to direct a lot of tax dollars over many years into understanding the molecular details of coronavirus spike proteins. The fact that only 5 percent of Africans were vaccinated against COVID-19 by the end of 2021 was not the result of any law of nature. It was the result of rich countries buying up the initial supplies, and of the world's failure to establish a global system for distributing them to poor countries.

Writing about science and politics is especially challenging, which is what makes Annie Lowrey's "American Motherhood" all the more impressive. In 2022, the United States Supreme Court ruled in *Dobbs v. Jackson Women's Health Organization* that abortion was no longer a right protected by the Constitution. That decision, ensured by Trump's appointment of three staunch opponents of abortion to the bench, has radically altered the options for millions of pregnant Americans. Sharing her own difficult experiences, Lowrey takes an unflinching look at the risks that come with pregnancy, and the threat that restrictions on reproductive freedom add to them.

Many pieces in this book address pressing news of the day. But the scope of science and nature writing is wide enough to include more than deadline reporting. History is rife with surprises and revelations. In "Another Green World" Jessica Camille Aguirre looks back at decades of obsession with the dream of building a self-sustaining biosphere in which we could hide from our self-inflicted woes. Ben Mauk turns our attention to an Asian rodent known as the marmot in "Shadows, Tokens, Spring." He uses this animal as the basis for a lyrical reflection on how animals have brought us many diseases before bats supplied the progenitors of SARS-CoV-2.

Sometimes what's most compelling about a scientific story is the way it challenges us to think about the concepts we take for granted. Science is no more insulated from philosophy than it is from politics. In "The Butterfly Effect," Maggie Koerth offers the sad tale of a species of butterfly close to extinction. But she also shows how its rescue will require us to think carefully about what it means for a species to survive in a human-dominated world, and why it should matter to us. In "Brain Wave," Ferris Jabr uses new advances in brain-machine interfaces to ask what will happen when brains and machines become inextricably linked. Where will people stop and the computers begin?

In any anthology like this, there needs to be room for straight-forward delight. It's always a pleasure to see Natalie Wolchover dive into the hardest questions about how the universe works. "A Deepening Crisis Forces Physicists to Rethink Structure of Nature's Laws" is a classic of this particular genre. The fact that thousands of fireflies can flash in synchrony makes the world a better place to live in. In "Bright Flight," Vanessa Gregory tells the story of scientists attempting to figure out how the insects achieve this feat with brains the size of peppercorns. The fact that some cows managed to survive a hurricane in 2019 is the antithesis of breaking news. And yet, in "True Grit," J. B. MacKinnon turns their odyssey into a story I couldn't put down. It even gives me a newfound respect for cows. Sabrina Imbler is a master at discovering the intersections between the biological world and our first-person experiences. "My Metamorphosis" is the shortest selection in this book. But calling Imbler's piece short is like calling a laser beam narrow. It's no less potent for the size.

While it's an honor to serve as a guest editor for *The Best American Science and Nature Writing*, I also consider it a way to pay back an old debt. I've been teaching science and nature writing to college students since 2009. Every year, I assign some of my favorite works, such as *Silent Spring* and magazine pieces including "Letting Go," by Atul Gawande, and "The Really Big One," by Kathryn Schulz. I also assign the latest edition of this anthology. Year after year, I have relied on other writers to curate a collection of recent work.

When I was approached about this year's edition, it seemed only fair for me to take a turn at shouldering the load.

Once I started looking back over 2022, I became both enthralled and overwhelmed. The collective effort of writers over the course of a year leaves us with so much to read. Every day, new reports come out about the torrent of scientific research. Investigative reporters dig into records to expose malfeasance, from hospitals turning away poor patients to corporations greenwashing their ongoing assault on the climate. Scientists publish first-person essays about their experiences carrying out research. Some of these pieces are published by magazines dedicated to science and nature. But many pop up far away on the journalistic landscape, like volunteer maples. A business reporter may examine how artificial intelligence threatens to turn Hollywood upside down. A sportswriter may tell a story about how the tradition of dogsled racing is disappearing in the warming Arctic. Great narratives about science and nature do not have to be typeset on paper. The words may float over a striking online visualization of data. They may buzz in earphones, creating the narration for an epic podcast.

I got my bearings again by thinking as a teacher. I try to teach the kind of class that I wish I had taken when I was young. Back then, I had no idea I would make science writing the work of my life. As a kid, I liked to write for fun, but I didn't produce articles about quantum teleportation. My output was a jumble of short stories, cartoons, and failed imitations of *Watership Down*. In college, I majored in English to learn from great writers while trying to avoid getting sucked into the self-annihilating maze of literary theory. I wrote short stories and spent a summer working at my county newspaper, reporting on local droughts and forgotten baseball leagues.

After college, I spent a couple of years at odd jobs—construction crew, administrative assistant—while trying to start a novel. I ended up getting hired an assistant copyeditor at a science magazine called *Discover*. At first it was just a day job to me. But I became so intrigued by what I was reading that I kept overlooking the missing commas and misspelled words.

My bosses looked kindly on me and let me switch to fact-checking, and then to writing short pieces of my own. This was, I gradually realized, a kind of writing fundamentally different from what I had done before. In nature, I was discovering strangeness beyond my own imagining. And scientists were willing to help me understand their discoveries, in long conversations over the phone or visits to their labs and field sites. I left fiction behind because reality kept surprising me.

As a teacher, I have tried to make explicit some of the rules I learned about what makes for good writing about science and nature. Shun jargon, embrace metaphor. Strive for zero uses of the passive voice. Make sure your paragraphs fit in a particular order. Bring people to life with visual details and judiciously selected quotes. Remember that the reader can read your story, not your mind.

There's a risk to providing too many rules: it may hem in the creativity of my students. If they become professional writers, they will have to deal with all sorts of constraints on their work, some necessary and some arbitrary. There will be style sheets. There will be inverted pyramids. But while they are still getting to know this particular genre, I want them to see how widely they can rove, how many ways there are to tell a story.

That is why I did not treat picking stories for this book like a baseball playoff. I would not try to exalt a few pieces over many other excellent ones. I decided instead to build this anthology as a teacher. I would create the best assortment of exemplary writing I could assemble—a selection that I would enjoy teaching. I suspect my students will enjoy this assignment.

CARL ZIMMER

The Best
AMERICAN
SCIENCE &
NATURE
WRITING
2023

ELIZABETH SVOBODA

An Invisible Epidemic

FROM *Scientific American*

IN EARLY 2021 emergency room physician Torree McGowan hoped the worst of the pandemic was behind her. She and her colleagues had adapted to the COVID-causing virus, donning layers of protection before seeing each patient, but they'd managed to keep things running smoothly. The central Oregon region where McGowan lived—a high desert plateau ringed by snow-capped mountains—had largely escaped the first COVID waves that slammed areas such as New York City.

Then the virus's Delta variant hit central Oregon with exponential fury, and the delicate balance McGowan had maintained came crashing down. Suddenly, COVID patients were streaming into the ERs at the hospitals where she worked, and she had to tell many patients she was powerless to help them because the few drugs she had didn't work in late stages of the disease. "That feels really terrible," McGowan says. "That's not what any of us signed up for."

It wasn't just COVID patients McGowan couldn't help. It was also everyone else. People still approached a health-care emergency with the expectation that they were going to be taken care of right away. But in the midst of the surge, there were no beds. "And I don't have a helicopter that can fly you between my hospital and the next hospital," she says, "because they're all full." A patient with suspected colon cancer showed up bleeding in the ER, and McGowan's inner impulses screamed that she needed to admit the woman immediately for testing. But because there were no beds left, she had to send the patient home instead.

The need to abandon her own standards and watch people suffer and die was hard enough for McGowan. Just as disorienting, though, was the sense that more and more patients no longer cared what happened to her or anyone else. She had assumed she and her patients played by the same basic rules—that she would try her utmost to help them get better and that they would support her or at least treat her humanely.

But as the virus extended its reach, those relationships broke down. Unvaccinated COVID patients walked into the exam room maskless, against hospital policy. They cursed her out for telling them they had the virus. "I have heard so many people say, 'I don't care if I make someone sick and kill them,'" McGowan says. Their ruthlessness simultaneously terrified and enraged her—not least because she had an immunocompromised husband at home. "Every month I do hours and hours of continuing education," McGowan says. "Every patient that I've ever made a mistake on, I can tell you every bit about that. And the thought that people are so callous with a life, when I place so much value on somebody's life—it's a lot to carry."

Moral injury is a specific trauma that arises when people face situations that deeply violate their conscience or threaten their core values. Those who grapple with it, such as McGowan, can struggle with guilt, anger and a consuming sense that they can't forgive themselves or others.

The condition affects millions across many roles. In an atmosphere of rationed care, doctors must admit a few patients and turn many away. Soldiers kill civilians to complete assigned missions. Veterinarians must put animals down when no one steps up to adopt them.

The trauma is far more widespread and devastating than most people realize. "It's really clear to us that it is all over the place," says psychiatrist Wendy Dean, president and co-founder of the nonprofit Moral Injury of Healthcare in Carlisle, Pennsylvania. "It's social workers, educators, lawyers." Survey studies in the US report that more than half of K-12 professionals, including teachers, moderately or strongly agree that they have faced morally injurious situations involving others. Similar studies in Europe show that about half of physicians have been exposed to potentially morally injurious events at high levels

Even these figures may be artificially low, given scant public awareness of moral injury: many people do not yet have the vocabulary to describe what is happening to them. Whatever the exact numbers, the mental health effects are vast. In a King's College London meta-analysis that surveyed thirteen studies, moral injury predicted higher rates of depression and suicidal impulses.

When COVID swept the planet, the moral injury crisis became more pressing as ethically wrenching dilemmas became the new normal—not just for health care workers but for others in frontline roles. Store employees had to risk their own safety and that of vulnerable family members to make a living. Lawyers often could not meet clients in person, making it nearly impossible to represent those clients adequately. In such situations, "no matter how hard you work, you're always going to be falling short," says California public defender Jenny Andrews.

Although moral injury doesn't yet have its own listing in diagnostic manuals, there is a growing consensus that it is a condition that is distinct from depression or post-traumatic stress disorder (PTSD). This consensus has given rise to treatments that aim to help people resolve long-standing ethical traumas. These treatments—vital additions to a broad range of trauma therapies—encourage people to face moral conflicts head on rather than blotting them out or explaining them away, and they emphasize the importance of community support in long-term recovery. In some cases, therapy clients even create plans to make amends for harms committed.

Even if moral injury research is a young and growing field, scientists and clinicians already agree that a key step toward healing for morally injured people—whether in therapy or not—has to do with grasping the true nature of what they're facing. They're not hopeless, "bad seeds" or uniquely irredeemable. They may not fit the criteria for PTSD or another mental illness. Instead they're suffering from a severe disconnect between the moral principles they live by and the reality of what is happening or has happened. In moral injury, "that sense of who you are as a person has been brought into question," Dean says. "We have a lot of people saying, 'This is the language I've been looking for for the past twenty years.'"

*

Ancient Origins

Although VA psychiatrist Jonathan Shay coined the term "moral injury" in the 1990s, the phenomenon predates its naming by millennia. In the ancient Greek epic *The Iliad*, the hero Achilles loses his best friend Patroclus in battle and then inwardly tortures himself because he failed to shield Patroclus from harm. When world wars broke out in the twentieth century, people labeled as "battle fatigued" the returning soldiers who bore mental scars. In reality, many of them were tortured not by shell shock but by wartime deeds they felt too ashamed to recount. In the 1980s, University of Nebraska Medical Center ethicist Andrew Jameton observed that this kind of moral distress was not confined to the military realm. It often "arises when one knows the right thing to do," he wrote, "but constraints make it nearly impossible to pursue the right course of action."

What spurred the first rigorous study of moral injury, however, was the multitude of US soldiers struggling after serving in wars in Vietnam, Iraq, and Afghanistan. Psychologist Brett Litz of the Veterans Affairs Boston Healthcare System saw quite a few vets of these conflicts who weren't responding well to counseling after their deployments ended. They seemed to be stuck in stagnant grief over acts they'd committed, such as killing civilians in war zones. They reminded Litz of one of his past therapists who'd seemed oddly detached, never mentally present in the room. Afterward Litz found out why, "Probably months before I went to him, he had opened his car door, and he killed a child who was just biking down the road," Litz says. "He was as broken as can be. I witnessed firsthand what that was."

In long conversations with veterans, Litz grew convinced he was witnessing a condition that was different from PTSD and depression. PTSD typically takes root when someone's life or safety is threatened. But much of the lingering trauma Litz saw in vets had nothing to do with direct personal threat. It was related to mounting guilt and hopelessness, "the totality of the inhumanity, the lack of meaning, and the participation in grotesque war things," he says. "They were pariahs—or felt that way, at least."

Building on Shay's earlier work, Litz resolved to develop a working concept of moral injury so that researchers could study it in depth and figure out how best to treat it. "I thought, 'This is going to affect our culture, and there are going to be broad impacts,'" he says. "We needed to bring science to bear. We needed to define the terms."

To that end, Litz and his colleagues published a comprehensive paper on moral injury in 2009, outlining common moral struggles veterans were facing and proposing a treatment approach that involved making personally meaningful reparations for harm done. He noted, too, that not all "potentially morally injurious events" cause moral injury. If you kill someone, and you feel totally justified in having done so, you may not experience moral injury at all. Moral injury tends to turn up when you have a vision of the world as fundamentally fair and good and something you've done or witnessed destroys that vision.

Litz's paper soon caught the attention of Rita Nakashima Brock, then a visiting scholar at Starr King School for the Ministry in California. A theologian and antiwar activist, Brock was preparing to convene the Commission on Conscience in War, an event where returned soldiers would testify about the moral impact of engaging in battle.

Brock's antiwar activism had personal roots. After her father, a US Army medic, returned from Vietnam, he withdrew from his family. When he did speak to his loved ones, he lashed out with an escalating rage. "My dad was so different that I didn't even want to be at home anymore," she says. After Brock's father died, she pieced together more of his story with a cousin's help. He had worked with a guide while deployed, a young Vietnamese woman who was later tortured and killed. He was horrified at what had happened—and likely also racked with guilt because he knew his ties to the guide could have put her in danger.

As soon as Brock saw Litz's moral injury paper, something clicked. "When my colleague and I read it, we said, 'Oh, my God, this is what the whole thing is about,'" she recalls. "We sent it to everybody testifying and said, 'Read this.'"

Chronicling the Unspeakable

After Brock's 2010 Truth Commission, her committee set forth a key objective: creating programs to inform the public about moral injury. With a grant from the private Lilly Endowment, Brock established a moral injury research and education program at Texas's Brite Divinity School. Later Tommy Potter—then a development officer at Brite—mentioned Brock's work to his childhood friend Mike King, CEO of the national nonprofit Volunteers of America (VOA), and Brock and King arranged a time to meet.

VOA had long focused on helping marginalized populations, and when Brock described the moral injury concept to King, "it just instantly resonated with every area of our work," King says. "It is profoundly there with veterans. But I could see it in our work with folks coming out of incarceration and certainly with health care." So in 2017 VOA put up about $1.3 million in funding to create the Shay Moral Injury Center in Alexandria, Va., named for Jonathan Shay. Brock became the center's first director, heading up research and training programs aimed at understanding and treating moral injury.

Meanwhile moral injury research at Litz's lab and elsewhere was starting to take flight. In 2013, along with his health-care colleagues, Litz debuted and road tested what he called the Moral Injury Events Scale, a measure of exposure to events that can cause moral injury. The scale assessed things such as how much people felt they'd violated their morals, how much they felt others had betrayed important values, and the level of distress they experienced as a result. Other investigators have confirmed moral injury can come with significant mental health burdens: in a 2019 study of five VA clinics across the US, people who'd experienced moral injury consistently had a higher risk of suicide than control participants.

Other research also backs up Litz's initial hunch that moral injury is distinct from PTSD, although the two conditions sometimes overlap. A 2019 study by researchers at the Salisbury VA Healthcare System in North Carolina reports that moral injury has different brain signatures than for PTSD alone: People with moral injury have more activity in the brain's precuneus area, which helps to govern moral judgments, than those who only have PTSD. And after people suffer moral traumas, they display different brain glucose metabolism patterns than those who suffer direct physical threats, according to a 2016 study by researchers at the University of Texas Health Science Center at San Antonio and their colleagues. The results support developing theories that moral injury is a unique biological entity.

As Brock's Shay Moral Injury Center was being established, she forged connections with powerful people who could get the word out about moral injury—including Margaret Kibben, the current chaplain at the US House of Representatives. Kibben holds regular events for House members, and one of her recent talks was about moral injury. The event drew about three times more members than usual, Brock reports, "and they all wanted to talk about their experience." Brock and Kibben's partnership reflects a growing trend in the study of moral injury: collaboration between scholars and clergy members who aim to chronicle the unspeakable and to help people through it. Moral injury "does really bring together a lot of disciplines," says psychologist Anna Harwood-Gross of Metiv, the Israel Psychotrauma Center in Jerusalem. "It's rare to see articles written by chaplains and psychologists together."

As COVID ravaged the planet from 2020 onward, moral injury research and inquiry took a distinct new turn. Health-care workers spoke out about how rationing care was affecting them psychologically, and Dean and her colleagues Breanne Jacobs and Rita Manfredi, both at the George Washington University School of Medicine and Health Sciences, published a journal article that urged employers to monitor moral injury's effects. "We need time, energy and intellectual capacity to make peace with those specters," they wrote.

The moral injury Dean sees in health care often doesn't stem from one-time, cataclysmic events. Many providers are suffering what she calls "death by a thousand cuts"—the constant, stultifying knowledge that they have to give people subpar care or none at all. "They think they suck. They think they're inadequate," says trauma surgeon Gregory Peck of New Jersey's Rutgers Robert Wood Johnson Medical School. "No one's putting their finger on 'You don't suck. This is moral injury you're suffering.'" Psychiatrist Mona Masood, who founded the Physician Support Line in 2020, has heard countless doctors agonize over daily moral compromises. "We'll hear, 'Am I really a failure? Have I failed my calling? Am I something not human anymore?'"

Off-Axis

Those words would surely resonate with McGowan. During an increase in COVID cases, as we approach a hospital where she works regular shifts, an ambulance pulls out of the parking lot, lights flashing, underneath looming clouds. "That's probably another transfer," McGowan says. Someone, in other words, has claimed one of the few available COVID beds in the region, meaning someone else—someone just as sick—may have to do without. Inside the ER, a bare-bones suite of rooms and hallways, glove boxes and black cords dangle from the walls. As we walk around, we spot warning signs of other moral compromises ahead. A scrawled note on a hallway whiteboard reads, "Critical shortage of green top tubes, 0-day supply of blue tops." When these tubes run out, McGowan explains, she may not be able to order blood tests patients need—and, as a result, may have a hard time figuring out what's wrong with them.

On many days during the pandemic, McGowan has struggled with the dislocation of shuttling between the ER—a personal hell of COVID deniers, irate family members and dying patients and the outside world, which feels disturbingly normal. How, she wonders, can people nonchalantly chat and sip coffee when, minutes before, she sent someone home who could barely breathe? How can her own moral world be knocked so profoundly off-axis while the larger world continues to spin with scarcely a wobble?

McGowan sees a therapist to help her process the situations she's faced, which she says has been helpful. Yet she continues to grapple with the fallout of moral dilemmas, reflecting a growing consensus that traditional therapy may not always be enough to help morally injured people get past lingering demons. Those who seek help sometimes make headway with basic cognitive-behavioral therapy (CBT), the current gold standard among insurers. Some researchers think CBT approaches are sufficient to treat moral injury.

But one sticking point with CBT is that it focuses on correcting clients' distorted thought patterns. For people with moral injury who've experienced wrenching events that upend their entire value system, ethical distress is genuine, not the product of distorted thinking, Harwood-Gross says. If people with moral injury simply try to retrain their thoughts, they may be left unsatisfied and unhealed.

Therapies for PTSD can likewise fall short for morally injured patients, in Harwood-Gross's experience. PTSD-focused approaches teach clients to adapt to traumatic triggers, such as fireworks that sound like gunshots, but this exposure approach doesn't really help them resolve deep ethical conflicts. Effective moral injury counseling is "more about the processing," Harwood-Gross says. "There has to be that movement: 'How do I see it for what it is and, from there, develop something more meaningful?' It's a more spiritual approach."

Recognizing moral injury's unique challenges, psychologists such as Litz have been creating therapies that more directly address clients' needs. Litz and other providers have pioneered a moral injury treatment called adaptive disclosure. Researchers at Australia's La Trobe University and University of Queensland have developed a similar approach called pastoral narrative disclosure. The latter involves discussing moral issues with a chaplain or other spiritual adviser rather than a doctor.

These therapies stress the importance of moral reckoning. They encourage clients to accept uncomfortable truths: "I led that attack on Iraqi civilians"; "I sent that suffering patient home without treatment." Then, with clients' input, counselors can help them develop strategies for making amends or pursuing closure say, apologizing to a family whose child they injured.

Early evidence suggests these approaches make headway where others can't. In Litz's initial trial of adaptive disclosure on forty-four marines, participants' negative beliefs about both themselves and the world diminished. Most also said the therapy helped to resolve their moral dilemmas.

Earlier this year Litz wrapped up a 173-person clinical trial of adaptive disclosure at VA sites in Boston, Minneapolis, San Diego, San Francisco, and central Texas. The trial's results have not yet been published, but Litz found that, in general, adaptive disclosure boosted participants' level of functioning over time. Litz says his goal is not to wipe people's moral slates clean but to restore their ability to thrive. "You'll never not feel awful when you think about what happened," Litz adds. "That's going to be the new normal. The question is 'How do you rehabilitate and live a good-enough life?'"

For Brock's VOA team, moral injury rehabilitation also involves a suite of peer-support programs. The Shay Moral Injury Center's core group offering, Resilience Strength Training (RST), is a sixty-hour, in-person program where people with moral injury share stories about events that caused it, engage in talks about forgiveness (for themselves or others), and do exercises to help them define their value system and purpose going forward In a survey study at two VOA program sites, participants scored an average of 46 percent higher on a scale of post-traumatic growth and 19 percent higher on a scale of perceived meaning in life than they had before starting the program. Although the in-person program paused during the pandemic, plans to restart it are currently underway.

In 2020 VOA created an online version of RST for health workers and others called Resilience Strength Time (ReST). Free ReST sessions run a few times a week, and attendees sign up for as many as they want.

During a recent ReST video meeting, several people showed up to talk for an hour about their moral challenges on the health-care front lines. One spoke about feeling helpless as she watched a patient verbally abuse a nurse giving vaccines. Peer-session leaders Bruce Gonseth and Jim Wong, both war vets, listened closely to each attendee's dilemma and empathized, often sharing recollections of similar situations they had faced. "To me, what we experienced in the war was exactly what frontline workers are experiencing: the invisible enemy," Wong told the group. "You may feel like you're letting other people down. You may observe others engaging in harmful behaviors. You're not alone. We're here to support you."

In most therapeutic relationships, a power differential exists between therapist and client. VOA's groups, where members and facilitators take turns being vulnerable, put participants on a more even footing. This openness builds bonds that support people's recovery, ensuring their moral struggles won't isolate them. "These are people who know them well and intimately, and it matters," Brock says. "Moral injury is a relationship break—you have an identity crisis. You have to establish new relationships that sustain you."

Therapies addressing moral injury that bolster clients' sense of purpose share a common goal with treatments developed by Austrian psychologist Viktor Frankl, who believed that a personal search for meaning could fuel trauma recovery. To survive his imprisonment in Nazi camps, including Auschwitz, Frankl focused on what motivated him to go on—his boundless love for his wife, his commitment to rewrite a research manuscript the Nazis had destroyed. "Everything can be taken from a man but one thing," Frankl wrote, "the last of the human freedoms—to choose one's attitude in any given set of circumstances, to choose one's own way." After his liberation in 1945, Frankl refined a treatment approach called logotherapy, which emphasized that a sense of purpose could help people endure the gravest suffering.

As Frankl would have done, therapists such as Litz and Harwood Gross encourage clients to accept the depth of inhumanity in the world rather than attempt to blot out awareness of that inhumanity. The essential question—the same one Frankl confronted—then becomes: "In the midst of what has happened and what is still happening, how can I find meaning in life?"

Partnerships between clinicians and religious leaders have helped facilitate that search for meaning, Brock says. Mental health treatment can feel like a formalized setup in which "the role of the professional is not to be personal," she says. But clergy often excel at connecting on a more informal, human level—an asset in dealing with morally injured people who have come to doubt their own humanity. "Chaplains don't bill by the hour," Brock says. "They spend the time they need to spend with people."

No Easy Way Out

Moral injury treatments are a needed safety valve for people battling guilt and ethical vertigo. Even so, as old hands on the front lines note, nudging the morally injured toward self-repair goes only so far. Therapy can help you move on from past choices, but unless your employer hires more staff or supplies more resources, chances are you'll have to keep making decisions that violate your ethics, compounding your trauma. A lot of problems that cause moral injury "require systemic solutions on a much broader level," says Andrews, the California public defender.

Yet many organizations are taking the easy way out, Dean says. Instead of launching systemic reforms that could help head off moral injury, they're offering "wellness solutions" such as massages and meditation tips, which can amount to putting a Band-Aid on a broken bone. "If I have to listen to another 'eat well, sleep well, do yoga' conversation, I'm going to throw up," says New York City ER doctor Jane Kim. What would be better, she thinks, is in-depth, system-wide conversations about what frontline workers actually need to do their jobs ethically, not what outside wellness providers assume they need. She argues that reforms based on these frank internal assessments would benefit both workers and those they serve. "We care for other people," she says. "But if we are broken ourselves, how can we possibly help others?"

As the pandemic drags on, similar thoughts occupy McGowan's mind. Although COVID hospital admissions have decreased somewhat in her area, workers have been quitting in droves, which means there still aren't enough providers to give patients adequate treatment. "I compare it to the Bataan Death March. There's no end in sight," McGowan says. On a bookshelf in her light-filled farmhouse, a plaque reads, "You never know how strong you are until being strong is the only choice you have."

The windows overlook a parched expanse of field. McGowan's farmer husband grew just a fraction of his usual hay crop last year because of a drought. In some ways, they face the same existential dilemma: What do you do when forces beyond your control shrivel your highest intentions?

To counter thoughts of hopelessness, of failing her medical calling, McGowan tries to focus on specific acts of good she's been able to perform. When she's not in the ER, she serves as a lieutenant colonel in the Oregon Air National Guard, and her unit has vaccinated more than a hundred thousand people against COVID.

Mentoring other doctors, too—offering advice as they process the same kinds of harrowing choices and regrets she's had—has buoyed her. "That's helped me learn to be a little bit kinder to myself," McGowan says. "The same words that I tell them, I try to repeat to myself: You did the best that you could." She inhales, hesitating. "And you are still a good doctor. I would still let you take care of my family."

SARAH ZHANG

The Myopia Generation

FROM *The Atlantic*

A DECADE INTO her optometry career, Marina Su began noticing something unusual about the kids in her New York City practice. More of them were requiring glasses, and at younger and younger ages. Many of these kids had parents who had perfect vision and who were baffled by the decline in their children's eyesight. Frankly, Su couldn't explain it either.

In optometry school, she had been taught—as American textbooks had been teaching for decades—that nearsightedness, or myopia, is a genetic condition. Having one parent with myopia doubles the odds that a kid will need glasses. Having two parents with myopia quintuples them. Over the years, she did indeed diagnose lots of nearsighted kids with nearsighted parents. These parents, she told me, would sigh in recognition: *Oh no, not them too.* But something was changing. A generation of children was suddenly seeing worse than their parents. Su remembers asking herself, as she saw more and more young patients with bad eyesight that seemed to have come out of nowhere: "If it's only genetics, then why are these kids also getting myopic?"

What she noticed in her New York office a few years ago has in fact been happening around the world. In East and Southeast Asia, where this shift is most dramatic, the proportion of teenagers and young adults with myopia has jumped from roughly a quarter to more than 80 percent in just over half a century. In China, myopia is so prevalent that it has become a national-security concern: The military is worried about recruiting enough sharp-eyed pilots from among the country's 1.4 billion people. Recent pandemic lockdowns seem to have made eyesight among Chinese children even worse.

For years, many experts dismissed the rising myopia rates in Asia as an aberration. They argued that Asians are genetically predisposed to myopia and nitpicked the methodology of studies conducted there. But eventually the scope of the problem and the speed of change became impossible to deny.

In the US, 42 percent of twelve- to fifty-four-year-olds were nearsighted in the early 2000s—the last time a national survey of myopia was conducted—up from a quarter in the 1970s. Though more recent large-scale surveys are not available, when I asked eye doctors around the US if they were seeing more nearsighted kids, the answers were: "Absolutely." "Yes." "No question about it."

In Europe as well, young adults are more likely to need glasses for distance vision than their parents or grandparents are now. Some of the lowest rates of myopia are in developing countries in Africa and South America. But where Asia was once seen as an outlier, it's now considered a harbinger. If current trends continue, one study estimates, half of the world's population will be myopic by 2050.

The consequences of this trend are more dire than a surge in bespectacled kids. Nearsighted eyes become prone to serious problems like glaucoma and retinal detachment in middle age, conditions that can in turn cause permanent blindness. The risks start small but rise exponentially with higher prescriptions. The younger myopia starts, the worse the outlook. In 2019, the American Academy of Ophthalmology convened a task force to recognize myopia as an urgent global-health problem. As Michael Repka, an ophthalmology professor at Johns Hopkins University and the AAO's medical director for government affairs, told me, "You're trying to head off an epidemic of blindness that's decades down the road."

The cause of this remarkable deterioration in our vision may seem obvious: You need only look around to see countless kids absorbed in phones and tablets and laptops. And you wouldn't be the first to conclude that staring at something inches from your face is bad for distance vision. Four centuries ago, the German astronomer Johannes Kepler blamed his own poor eyesight, in part, on all the hours he spent studying. Historically, British doctors have found myopia to be much more common among Oxford students than among military recruits, and in "more rigorous" town schools than in rural ones. A late nineteenth-century ophthalmology handbook even suggested treating myopia with a change of air and avoidance of all work with the eyes—"a sea voyage if possible."

By the early twentieth century, experts were coalescing around the idea that myopia was caused by "near work," which might include reading and writing—or, these days, watching TV and scrolling through Instagram. In China, officials have become so alarmed that they've proposed large-scale social changes to curb myopia in children. Written exams are now limited before third grade, and video games are restricted. One elementary school reportedly installed metal bars on its desks to prevent kids from leaning in too close to their schoolwork.

Spend too much time scrutinizing text or images right in front of you, the logic goes, and your eyes become nearsighted. "Long ago, humans were hunters and gatherers," says Liandra Jung, an optometrist in the Bay Area. We relied on our sharp distance vision to track prey and find ripe fruit. Now our modern lives are close-up and indoors. "To get food, we forage by getting Uber Eats."

This is a pleasingly intuitive explanation, but it has been surprisingly difficult to prove. "For every study that shows an effect of near work on myopia, there's another study that doesn't," says Thomas Aller, an optometrist in San Bruno, California. Adding up the number of hours spent in front of a book or screen does not seem to explain the onset or progression of nearsightedness.

A number of theories have rushed to fill this confusing vacuum. Maybe the data in the studies are wrong—participants didn't record their hours of near work accurately. Maybe the total duration of near work is less important than whether it's interrupted by short breaks. Maybe it's not near work itself that ruins eyes but the fact that it deprives kids of time outdoors. Scientists who argue for the importance of the outdoors are further sub divided into two camps: those who believe that bright sunlight promotes proper eye growth versus those who believe that wide-open spaces do.

Something about modern life is destroying our ability to see far away, but what?

Asking this question will plunge you into a thicket of scientific rivalries—which is what happened when I asked Christine Wildsoet, an optometry professor at UC Berkeley, about the biological plausibility of these myopia theories. Over the course of two hours, she paused repeatedly to note that the next part was contentious. "I'm not sure which controversy we're up to," she said at one point. (It was no. 4, and there were still three more to come.) But, she also noted, these theories are essentially two sides of the same coin: Anyone who does too much near work is also not spending much time outside. Whichever theory is true, you can draw the same practical conclusion about what's best for kids' vision: less time hunched over screens, more time on outdoor activities.

By now, scientists have moved past the faulty assumption that myopia is purely genetic. That idea took hold in the '60s, when studies of twins showed that identical twins had more similar patterns of myopia than fraternal ones, and persisted in the academic world for decades. DNA does indeed play a role in myopia, but the tricky factor here is that identical twins don't just share the same genes; they're exposed to many of the same environmental stimuli, too.

Glasses, contacts, and laser surgery all help nearsighted people see better. But none of these fixes corrects the underlying anatomical problem of myopia. Whereas a healthy eye is shaped almost like an orb, a nearsighted one is more like an olive. To slow the progression of myopia, we would have to stop the elongation of the eyeball.

Which we already know how to do. Treatments to slow the progression of myopia—called "myopia control" or "myopia management"—exist. They're just not widely known in America.

Over the past two decades, eye doctors—mostly in Asia—have discovered that special lenses and eye drops can slow the progression of nearsightedness in children. Maria Liu, a myopia researcher who grew up in Beijing, told me that she first became interested in nearsightedness as a teenager, when she began watching classmates at her school for gifted children get glasses one by one. In this intensely competitive academic environment, she remembers spending the hours of 6:30 a.m. to 10 p.m. on schoolwork, virtually all indoors. By the time she finished university, nearly all of her fellow students needed glasses, and she did too.

Years later, when she started an ophthalmology residency in China, she met many young patients who wore orthokeratology lenses—also known as OrthoK—a type of overnight contact lens that temporarily alters the way light enters the eye by reshaping the clear front layer of the eyeball, thus improving vision during the day. Liu noticed, anecdotally, that those who wore OrthoK seemed to have better vision down the line than those who wore glasses. Could long-term use of the lenses somehow prevent elongation of the eye, thus impeding myopia's progression? It turns out that other scientists and doctors across Asia were noticing the same trend. In 2004, a randomized controlled study in Hong Kong of OrthoK confirmed Liu's hunch.

By then, Liu had moved to the US, and she soon began a doctoral program in vision science at Berkeley to study myopia. Her classmates, she recalls, were tackling exotic-sounding topics such as gene therapy and retinal transplants and wondered why she was studying "something that's so boring." She ended up working in Wildsoet's lab, researching the development of myopia in young chick eyes.

In humans, the majority of babies are born farsighted. Our eyes start slightly too short, and they grow in childhood to the right length, then stop. This process has been finely calibrated over millions of years of evolution. But when the environmental signals don't match what the eye has evolved to expect—whether that's due to too much near work, not enough outdoor time, some combination of the two, or another factor—the eye just keeps growing. This process is irreversible. "You can't make a longer eyeball shorter," Liu said. But you can interrupt growth by counteracting these faulty signals, which is what myopia control is designed to do.

When Liu became a professor at Berkeley after receiving her PhD, she started envisioning a myopia-control clinic—the first of its kind in the US—that could bridge the gap between research and practice. By then, she knew that many doctors in China were already successfully using OrthoK for myopia control.

The school administration was skeptical. Liu says that the clinical director didn't see how the clinic would benefit optometry students, or how it could attract enough patients to be worthwhile financially. But in 2013, Liu started it anyway, as a one-woman operation. She began seeing patients on Sundays in borrowed exam rooms with no extra pay and without relinquishing any of her teaching or clinical duties. Within months, her schedule was full. The Berkeley Myopia Control Clinic now runs four days a week and has a thousand active patients—some of whom drive hours through Bay Area traffic to get there. Liu was one of the only people at the school who anticipated the clinic's massive success. Jung, who is also an assistant clinical professor at Berkeley, told me that Liu's knowledge of the latest myopia-control treatments made it feel like she came "from the future."

When I arrived at the clinic at 8 a.m. on a Saturday morning this past spring—an hour at which the rest of the campus was still quiet—it was already filling up with optometry students and residents who work there as part of their training. Liu, who is petite with neat, wavy hair, moved through the clinic with frightful efficiency. One moment she was examining eyes, the next talking down a parent whose son's contact-lens shipment had gone missing, the next warning staffers about a malfunctioning printer.

The clinic offers three different treatments: OrthoK, multifocal soft contact lenses, and atropine eye drops. The first two both work by tweaking how light enters the eye, producing a signal for the eyeball to stop lengthening. Atropine, in contrast, is a drug that seems to chemically alter the growth pathway of the eye when used at low doses. (It also dilates the pupil; Cleopatra reportedly used it to make her eyes more beautiful.) These treatments slow myopia progression on average by about 50 percent. The original clinical trials validating them were mostly conducted in Asia starting in the mid-2000s. And the American Optometric Association's evidence-based committee published a report advising its members on how to use myopia control last year. Until quite recently, though, none of these treatments had been approved by the FDA for myopia control. Any optometrists who wanted to offer them had to go off label. And any patient who wanted to use them had to find the right doctor.

It's not a coincidence that Liu's clinic found early success in the Bay Area, which has a large Asian population. Eye doctors I spoke with in multiple cities across the US said it was usually Asian parents who came in asking for myopia control. The parents I met at the clinic skewed Asian and, on that Saturday, particularly Chinese—first-generation immigrants who speak Mandarin seek Liu out on the days she is personally in the clinic. Many of them heard about myopia control from fellow immigrants or friends in Asia. George Tsai, whose eight-year-old son was at the clinic for an OrthoK appointment, told me that his wife, who grew up in China, had learned of myopia control through WeChat, the messaging app popular in the country and among the Chinese diaspora.

Liu has a second phone, which she uses to manage three WeChat groups full of parents with kids in myopia control across North America. The questions flood in day and night. "First thing in the morning, I look at this WeChat group. Who has lost a lens? Who has red eyes? Who has other problems?" she said. "And again, before I go to bed." She started the first group with a parent of one of her patients. When it hit the maximum number of members allowed on WeChat, they created a second, and then a third. The groups now contain a total of fifteen hundred parents.

In general, Liu told me, Asian parents tend to be a lot more motivated because myopia "is much better perceived or accepted as a disease in Asian culture." I know this firsthand, as the child of Chinese immigrants. Distressed about my worsening vision in elementary school, my mother would regularly admonish me, standing my pencil case upright to measure the distance between my head and my desk. She also made me do eye exercises developed in China, which I was vindicated to finally learn, in the course of reporting this story, do not work. This was the late '90s, when there really was nothing to be done about myopia progression. But in the parents I met at the Berkeley clinic, I saw the same determination I once saw in my own. They had uprooted their lives and come to a foreign country and now here they were, hoping to bestow upon their kids any advantage, any edge that modern science could give.

There is another reason that the Bay Area, with its high median income, has been fertile ground for myopia control: The treatments are expensive. Many of the parents I met at the clinic were engineers or doctors. At Berkeley, OrthoK costs more than $450 for one pair of lenses, plus $1,600 for the initial fitting, not including the fees for several follow-up appointments a year. Soft contact lenses can run from several hundred to more than $1,000 a year. And a year's supply of atropine eye drops costs hundreds of dollars. Kids are typically in myopia control until their mid-teens to early twenties. Vision insurance does not cover any of these treatments.

Multinational eye-care companies now see myopia control as a hot potential market. They're vying for FDA approval of new lenses and improved formulations of atropine, which can be patented rather than sold as a cheaper generic. The business case is obvious: If half of the world is myopic by 2050, that's a huge pool of would-be customers. "How often do you have an opportunity to have an impact on a condition that will affect one out of two people? There's nothing else on the planet that I'm aware of," says Joe Rappon, the former chief medical officer of SightGlass Vision, a small California company whose myopia-control technology was jointly acquired by the eye-care giants CooperVision and Essilor.

In November 2019, the FDA green-lighted the first—and currently only—treatment specifically designed to slow the progression of myopia in the US, a soft contact lens from CooperVision called MiSight. Many more treatments, though, are in trials in the US, including several types of spectacles that tweak the way light enters the eye in order to slow its growth. Some are already on the market in Europe and Canada.

Once those glasses get approved in the US, "that's going to open the floodgates of myopia management," Barry Eiden, an optometrist in Deerfield, Illinois, told me. The earlier you can start slowing myopia progression in kids, the better the outcome, he explained, but parents sometimes balk at the idea of putting drugs or contacts into the eyes of their young children. They don't have the same problem with glasses.

In the future, Liu told me, she hopes FDA approvals will spur vision insurance to cover myopia control at least partially, making the treatments affordable to more parents. Meanwhile, CooperVision has already revved up its MiSight marketing machine. It's targeting exactly the parents you would expect: In my own Brooklyn neighborhood of Park Slope, where you regularly see toddlers in $1,000-plus Uppababy strollers, an optometry shop recently hung a big banner advertising MiSight with two smiling kids. An optometrist in downtown San Francisco told me that parents who have seen MiSight's ads are now coming into her office asking for it by name. The word-of-mouth era of myopia control is ending; the mass-advertising era is beginning.

Within the optometry business, myopia control often gets compared to braces—another treatment for which middle- and upper-class parents who want the best for their kids will dutifully shell out thousands of dollars. This comparison feels apt in a different way, too. Braces are also a modern solution to a relatively modern affliction. The teeth of cavemen, anthropologists have marveled, were incredibly straight. Crooked teeth appear in the archaeological record only when our ancestors transitioned from chewing raw meat and vegetables to eating cooked and processed grains. Our jaws are now smaller and weaker from disuse, our teeth more crowded and crooked. Today, braces are the way we retrofit our ill-adapted bodies for contemporary life.

We may not know exactly how ogling screens all day and spending so much time indoors are affecting us, or which is doing more damage, but we do know that myopia is a clear consequence of living at odds with our biology. The optometrists I spoke with all said they try to push better vision habits, such as limiting screen time and playing outside. But this goes only so far. Today, taking a phone away from a teenager may be no more practical than feeding a toddler a raw hunter-gatherer diet.

So this is where we've ended up, for those of us who can even afford it: adding chemicals and putting pieces of plastic in our eyes every day, in hopes of tricking them back to their natural state.

NATALIE WOLCHOVER

A Deepening Crisis Forces Physicists to Rethink Structure of Nature's Laws

FROM *Quanta Magazine*

IN *The Structure of Scientific Revolutions*, the philosopher of science Thomas Kuhn observed that scientists spend long periods taking small steps. They pose and solve puzzles while collectively interpreting all data within a fixed worldview or theoretical framework, which Kuhn called a paradigm. Sooner or later, though, facts crop up that clash with the reigning paradigm. Crisis ensues. The scientists wring their hands, reexamine their assumptions, and eventually make a revolutionary shift to a new paradigm, a radically different and truer understanding of nature. Then incremental progress resumes.

For several years, the particle physicists who study nature's fundamental building blocks have been in a textbook Kuhnian crisis.

The crisis became undeniable in 2016, when, despite a major upgrade, the Large Hadron Collider in Geneva still hadn't conjured up any of the new elementary particles that theorists had been expecting for decades. The swarm of additional particles would have solved a major puzzle about an already known one, the famed Higgs boson. The hierarchy problem, as the puzzle is called, asks why the Higgs boson is so lightweight—a hundred million billion times less massive than the highest energy scales that exist in nature. The Higgs mass seems unnaturally dialed down relative to these higher energies, as if huge numbers in the underlying equation that determines its value all miraculously cancel out.

The extra particles would have explained the tiny Higgs mass, restoring what physicists call "naturalness" to their equations. But after the LHC became the third and biggest collider to search in vain for them, it seemed that the very logic about what's natural in nature might be wrong. "We are confronted with the need to reconsider the guiding principles that have been used for decades to address the most fundamental questions about the physical world," Gian Giudice, head of the theory division at CERN, the lab that houses the LHC, wrote in 2017.

At first, the community despaired. "You could feel the pessimism," said Isabel Garcia Garcia, a particle theorist at the Kavli Institute for Theoretical Physics at the University of California, Santa Barbara, who was a graduate student at the time. Not only had the $10 billion proton smasher failed to answer a forty-year-old question, but the very beliefs and strategies that had long guided particle physics could no longer be trusted. People wondered more loudly than before whether the universe is simply unnatural, the product of fine-tuned mathematical cancellations. Perhaps there's a multiverse of universe, all with randomly dialed Higgs masses and other parameters, and we find ourselves here only because our universe's peculiar properties foster the formation of atoms, stars, and planets and therefore life. This "anthropic argument," though possibly right, is frustratingly untestable.

Many particle physicists migrated to other research areas, "where the puzzle hasn't gotten as hard as the hierarchy problem," said Nathaniel Craig, a theoretical physicist at UCSB.

Some of those who remained set to work scrutinizing decades-old assumptions. They started thinking anew about the striking features of nature that seem unnaturally fine-tuned—both the Higgs boson's small mass, and a seemingly unrelated case, one that concerns the unnaturally low energy of space itself. "The really fundamental problems are problems of naturalness," Garcia Garcia said.

Their introspection is bearing fruit. Researchers are increasingly zeroing in on what they see as a weakness in the conventional reasoning about naturalness. It rests on a seemingly benign assumption, one that has been baked into scientific outlooks since ancient Greece: Big stuff consists of smaller, more fundamental stuff—an idea known as reductionism. "The reductionist paradigm . . . is hard-wired into the naturalness problems," said Nima Arkani-Hamed, a theorist at the Institute for Advanced Study in Princeton, New Jersey.

Now a growing number of particle physicists think naturalness problems and the null results at the Large Hadron Collider might be tied to reductionism's breakdown. "Could it be that this changes the rules of the game?" Arkani-Hamed said. In a slew of recent papers, researchers have thrown reductionism to the wind. They're exploring novel ways in which big and small distance scales might conspire, producing values of parameters that look unnaturally fine-tuned from a reductionist perspective.

"Some people call it a crisis. That has a pessimistic vibe associated to it and I don't feel that way about it," said Garcia Garcia. "It's a time where I feel like we are on to something profound."

What Naturalness Is

The Large Hadron Collider did make one critical discovery: In 2012, it finally struck upon the Higgs boson, the keystone of the fifty-year-old set of equations known as the Standard Model of particle physics, which describes the seventeen known elementary particles.

The discovery of the Higgs confirmed a riveting story that's written in the Standard Model equations. Moments after the Big Bang, an entity that permeates space called the Higgs field suddenly became infused with energy. This Higgs field crackles with Higgs bosons, particles that possess mass because of the field's energy. As electrons, quarks, and other particles move through space, they interact with Higgs bosons, and in this way they acquire mass as well.

After the Standard Model was completed in 1975, its architects almost immediately noticed a problem.

When the Higgs gives other particles mass, they give it right back; the particle masses shake out together. Physicists can write an equation for the Higgs boson's mass that includes terms from each particle it interacts with. All the massive Standard Model particles contribute terms to the equation, but these aren't the only contributions. The Higgs should also mathematically mingle with heavier particles, up to and including phenomena at the Planck scale, an energy level associated with the quantum nature of gravity, black holes, and the Big Bang. Planck-scale phenomena should contribute terms to the Higgs mass that are huge—roughly a hundred million billion times larger than the actual Higgs mass. Naively, you would expect the Higgs boson to be as heavy as they are, thereby beefing up other elementary particles as well. Particles would be too heavy to form atoms, and the universe would be empty.

For the Higgs to depend on enormous energies yet end up so light, you have to assume that some of the Planckian contributions to its mass are negative while others are positive, and that they're all dialed to just the right amounts to exactly cancel out. Unless there's some reason for this cancellation, it seems ludicrous—about as unlikely as air currents and table vibrations counteracting each other to keep a pencil balanced on its tip. This kind of fine-tuned cancellation physicists deem "unnatural."

Within a few years, physicists found a tidy solution: supersymmetry, a hypothesized doubling of nature's elementary particles. Supersymmetry says that every boson (one of two types of particle) has a partner fermion (the other type), and vice versa. Bosons and fermions contribute positive and negative terms to the Higgs mass, respectively. So if these terms always come in pairs, they'll always cancel.

The search for supersymmetric partner particles began at the Large Electron-Positron Collider in the 1990s. Researchers assumed the particles were just a tad heavier than their Standard Model partners, requiring more raw energy to materialize, so they accelerated particles to nearly light speed, smashed them together, and looked for heavy apparitions among the debris.

Meanwhile, another naturalness problem surfaced.

The fabric of space, even when devoid of matter, seems as if it should sizzle with energy—the net activity of all the quantum fields coursing through it. When particle physicists add up all the presumptive contributions to the energy of space, they find that, as with the Higgs mass, injections of energy coming from Planck-scale phenomena should blow it up. Albert Einstein showed that the energy of space, which he dubbed the cosmological constant, has a gravitationally repulsive effect; it causes space to expand faster and faster. If space were infused with a Planckian density of energy, the universe would have ripped itself apart moments after the Big Bang. But this hasn't happened.

Instead, cosmologists observe that space's expansion is accelerating only slowly, indicating that the cosmological constant is small. Measurements in 1998 pegged its value as a million million million million million times lower than the Planck energy. Again, it seems all those enormous energy injections and extractions in the equation for the cosmological constant perfectly cancel out, leaving space eerily placid.

Both of these big naturalness problems were evident by the late 1970s, but for decades, physicists treated them as unrelated. "This was in the phase where people were schizophrenic about this," said Arkani-Hamed. The cosmological constant problem seemed potentially related to mysterious, quantum aspects of gravity, since the energy of space is detected solely through its gravitational effect. The hierarchy problem looked more like a "dirty-little-details problem," Arkani-Hamed said—the kind of issue that, like two or three other problems of the past, would ultimately reveal a few missing puzzle pieces. "The sickness of the Higgs," as Giudice called its unnatural lightness, was nothing a few supersymmetry particles at the LHC couldn't cure.

In hindsight, the two naturalness problems seem more like symptoms of a deeper issue.

"It's useful to think about how these problems come about," said Garcia Garcia in a Zoom call from Santa Barbara this winter. "The hierarchy problem and the cosmological constant problem are problems that arise in part because of the tools we're using to try to answer questions—the way we're trying to understand certain features of our universe."

Reductionism Made Precise

Physicists come by their funny way of tallying contributions to the Higgs mass and cosmological constant honestly. The calculation method reflects the strange nesting-doll structure of the natural world.

Zoom in on something, and you'll discover that it's actually a lot of smaller things. What looks from afar like a galaxy is really a collection of stars; each star is many atoms; an atom further dissolves into hierarchical layers of subatomic parts. Moreover, as you zoom in to shorter distance scales, you see heavier and more energetic elementary particles and phenomena—a profound link between high energies and short distances that explains why a high-energy particle collider acts like a microscope on the universe. The connection between high energies and short distances has many avatars throughout physics. For instance, quantum mechanics says every particle is also a wave; the more massive the particle, the shorter its associated wavelength. Another way to think about it is that energy has to cram together more densely to form smaller objects. Physicists refer to low-energy, long-distance physics as "the IR," and high-energy, short-distance physics as "the UV," drawing an analogy with infrared and ultraviolet wavelengths of light.

In the 1960s and '70s, the particle physics titans Kenneth Wilson and Steven Weinberg put their finger on what's so remarkable about nature's hierarchical structure: It allows us to describe goings-on at some big, IR scale of interest without knowing what's "really" happening at more microscopic, UV scales. You can, for instance, model water with a hydrodynamic equation that treats it as a smooth fluid, glossing over the complicated dynamics of its H_2O molecules. The hydrodynamic equation includes a term representing water's viscosity—a single number, which can be measured at IR scales, that summarizes all those molecular interactions happening in the UV. Physicists say IR and UV scales "decouple," which lets them effectively describe aspects of the world without knowing what's going on deep down at the Planck scale— the ultimate UV scale, corresponding to a billionth of a trillionth of a trillionth of a centimeter, or ten billion billion gigaelectronvolts (GeV) of energy, where the very fabric of space-time probably dissolves into something else.

"We can do physics because we can remain ignorant about what happens at short distances," said Riccardo Rattazzi, a theoretical physicist at the Swiss Federal Institute of Technology Lausanne.

Wilson and Weinberg separately developed pieces of the framework that particle physicists use to model different levels of our nesting-doll world: effective field theory. It's in the context of EFT that naturalness problems arise.

An EFT models a system—a bundle of protons and neutrons, say—over a certain range of scales. Zoom in on protons and neutrons for a while and they will keep looking like protons and neutrons; you can describe their dynamics over that range with "chiral effective field theory." But then an EFT will reach its "UV cutoff," a short-distance, high-energy scale at which the EFT stops being an effective description of the system. At a cutoff of 1 GeV, for example, chiral effective field theory stops working, because protons and neutrons stop behaving like single particles and instead act like trios of quarks. A different theory kicks in.

Importantly, an EFT breaks down at its UV cutoff for a reason. The cutoff is where new, higher-energy particles or phenomena that aren't included in that theory must be found.

In its range of operation, an EFT accounts for UV physics below the cutoff by adding "corrections" representing these unknown effects. It's just like how a fluid equation has a viscosity term to capture the net effect of short-distance molecular collisions. Physicists don't need to know what actual physics lies at the cutoff to write these corrections; they just use the cutoff scale as a ballpark estimate of the size of the effects.

Typically when you're calculating something at an IR scale of interest, the UV corrections are small, proportional to the (relatively smaller) length scale associated with the cutoff. The situation changes, though, when you're using EFT to calculate a parameter like the Higgs mass or the cosmological constant—something that has units of mass or energy. Then the UV corrections to the parameter are big, because (to have the right units) the corrections are proportional to the energy—rather than the length—associated with the cutoff. And while the length is small, the energy is high. Such parameters are said to be "UV-sensitive."

The concept of naturalness emerged in the 1970s along with effective field theory itself, as a strategy for identifying where an EFT must cut off, and where, therefore, new physics must lie. The logic goes like this: If a mass or energy parameter has a high cutoff, its value should naturally be large, pushed higher by all the UV corrections. Therefore, if the parameter is small, the cutoff energy must be low.

Some commentators have dismissed naturalness as a mere aesthetic preference. But others point to when the strategy revealed precise, hidden truths about nature. "The logic works," said Craig, a leader of recent efforts to rethink that logic. Naturalness problems "have always been a signpost of where the picture changes and new things should appear."

What Naturalness Can Do

In 1974, a few years before the term "naturalness" was even coined, Mary K. Gaillard and Ben Lee made spectacular use of the strategy to predict the mass of a then-hypothetical particle called the charm quark. "The success of her prediction and its relevance to the hierarchy problem are wildly underappreciated in our field," Craig said.

That summer of '74, Gaillard and Lee were puzzling over the difference between the masses of two kaon particles—composites of quarks. The measured difference was small. But when they tried to calculate this mass difference with an EFT equation, they saw that its value was at risk of blowing up. Because the kaon mass difference has units of mass, it's UV-sensitive, receiving high-energy corrections coming from the unknown physics at the cutoff. The theory's cutoff wasn't known, but physicists at the time reasoned that it couldn't be very high, or else the resulting kaon mass difference would seem curiously small relative to the corrections—unnatural, as physicists now say. Gaillard and Lee inferred their EFT's low cutoff scale, the place where new physics should reveal itself. They argued that a recently proposed quark called the charm quark must be found with a mass of no more than 1.5 GeV.

The charm quark showed up three months later, weighing 1.2 GeV. The discovery ushered in a renaissance of understanding known as the November revolution that quickly led to the completion of the Standard Model. In a recent video call, Gaillard, now eighty-two, recalled that she was in Europe visiting CERN when the news broke. Lee sent her a telegram: CHARM HAS BEEN FOUND.

Such triumphs led many physicists to feel certain that the hierarchy problem, too, should herald new particles not much heavier than those of the Standard Model. If the Standard Model's cutoff were up near the Planck scale (where researchers know for sure that the Standard Model fails, since it doesn't account for quantum gravity), then the UV corrections to the Higgs mass would be huge—making its lightness unnatural. A cutoff not far above the mass of the Higgs boson itself would make the Higgs about as heavy as the corrections coming from the cutoff, and everything would look natural. "That option has been the starting point of the work that has been done in trying to address the hierarchy problem in the last forty years," said Garcia Garcia. "People came up with great ideas, like supersymmetry, compositeness [of the Higgs], that we haven't seen realized in nature."

Garcia Garcia was a few years into her particle physics doctorate at the University of Oxford in 2016 when it became clear to her that a reckoning was in order. "That's when I became more interested in this missing component that we don't normally incorporate when we discuss these problems, which is gravity—this realization that there's more to quantum gravity than we can tell from effective field theory."

Gravity Mixes Everything Up

Theorists learned in the 1980s that gravity doesn't play by the usual reductionist rules. If you bash two particles together hard enough, their energies become so concentrated at the collision point that they'll form a black hole—a region of such extreme gravity that nothing can escape. Bash particles together even harder, and they'll form a bigger black hole. More energy no longer lets you see shorter distances—quite the opposite. The harder you bash, the bigger the resulting invisible region is. Black holes and the quantum gravity theory that describes their interiors completely reverse the usual relationship between high energies and short distances. "Gravity is anti-reductionist," said Sergei Dubovsky, a physicist at New York University.

Quantum gravity seems to toy with nature's architecture, making a mockery of the neat system of nested scales that EFT-wielding physicists have grown accustomed to. Craig, like Garcia Garcia, began to think about the implications of gravity soon after the LHC's search came up empty. In trying to brainstorm new solutions to the hierarchy problem, Craig reread a 2008 essay about naturalness by Giudice, the CERN theorist. He started wondering what Giudice meant when he wrote that the solution to the cosmological constant problem might involve "some complicated interplay between infrared and ultraviolet effects." If the IR and the UV have complicated interplay, that would defy the usual decoupling that allows effective field theory to work. "I just Googled things like 'UV-IR mixing,'" Craig said, which led him to some intriguing papers from 1999, "and off I went."

UV-IR mixing potentially resolves naturalness problems by breaking EFT's reductionist scheme. In EFT, naturalness problems arise when quantities like the Higgs mass and the cosmological constant are UV-sensitive, yet somehow don't blow up, as if there's a conspiracy between all the UV physics that nullifies their effect on the IR. "In the logic of effective field theory, we discard that possibility," Craig explained. Reductionism tells us that IR physics emerges from UV physics—that water's viscosity comes from its molecular dynamics, protons get their properties from their inner quarks, and explanations reveal themselves as you zoom in—never the reverse. The UV isn't influenced or explained by the IR, "so [UV effects] can't have a conspiracy to make things work out for the Higgs at a very different scale."

The question Craig now asks is: "Could that logic of effective field theory break down?" Perhaps explanations really can flow both ways between the UV and the IR. "That's not totally pie in the sky, because we know that gravity does that," he said. "Gravity violates the normal EFT reasoning because it mixes physics at all length scales—short distances, long distances. Because it does that, it gives you this way out."

How UV-IR Mixing Might Save Naturalness

Several new studies of UV-IR mixing and how it might solve naturalness problems refer back to two papers that appeared in 1999. "There is a growth of interest in these more exotic, non-EFT-like solutions to these problems," said Patrick Draper, a professor at the University of Illinois, Urbana-Champaign, whose recent work picks up where one of the 1999 papers left off.

Draper and his colleagues study the CKN bound, named for the authors of the '99 paper, Andrew Cohen, David B. Kaplan, and Ann Nelson. The authors thought about how, if you put particles in a box and heat it up, you can increase the energy of the particles only so much before the box collapses into a black hole. They calculated that the number of high-energy particle states you can fit in the box before it collapses is proportional to the box's surface area raised to the three-fourths power, not the box's volume, as you might think. They realized that this represented a strange UV-IR relationship. The size of the box, which sets the IR scale, severely limits the number of high-energy particle states within the box—the UV scale.

They then realized that if their same bound applies to our entire universe, it resolves the cosmological constant problem. In this scenario, the observable universe is like a very large box. And the number of high-energy particle states it can contain is proportional to the observable universe's surface area to the three-fourths power, not the universe's (much larger) volume.

That means the usual EFT calculation of the cosmological constant is too naive. That calculation tells the story that high-energy phenomena should appear when you zoom in on the fabric of space, and this should blow up the energy of space. But the CKN bound implies that there may be far, far less high-energy activity than the EFT calculation assumes—meaning precious few high-energy states available for particles to occupy. Cohen, Kaplan, and Nelson did a simple calculation showing that, for a box the size of our universe, their bound predicts more or less exactly the tiny value for the cosmological constant that's observed.

Their calculation implies that big and small scales might correlate with each other in a way that becomes apparent when you look at an IR property of the whole universe, such as the cosmological constant.

Draper and Nikita Blinov confirmed in another crude calculation last year that the CKN bound predicts the observed cosmological constant; they also showed that it does so without ruining the many successes of EFT in smaller-scale experiments.

The CKN bound doesn't tell you why the UV and IR are correlated—why, that is, the size of the box (the IR) severely limits the number of high-energy states within the box (the UV). For that, you probably need to know quantum gravity.

Other researchers have looked for answers in a specific theory of quantum gravity: string theory. Last summer, the string theorists Steven Abel and Keith Dienes showed how UV-IR mixing in string theory might address both the hierarchy and cosmological constant problems.

A candidate for the fundamental theory of gravity and everything else, string theory holds that all particles are, close up, little vibrating strings. Standard Model particles like photons and electrons are low-energy vibration modes of the fundamental string. But the string can wiggle more energetically as well, giving rise to an infinite spectrum of string states with ever-higher energies. The hierarchy problem, in this context, asks why corrections from these string states don't inflate the Higgs, if there's nothing like supersymmetry to protect it.

Dienes and Abel calculated that, because of a different symmetry of string theory called modular invariance, corrections from string states at all energies in the infinite spectrum from IR to UV will be correlated in just the right way to cancel out, keeping both the Higgs mass and the cosmological constant small. The researchers noted that this conspiracy between low- and high-energy string states doesn't explain why the Higgs mass and the Planck energy are so widely separated to begin with, only that such a separation is stable. Still, in Craig's opinion, "it's a really good idea."

The new models represent a growing grab bag of UV-IR mixing ideas. Craig's angle of attack traces back to the other 1999 paper, by the prominent theorist Nathan Seiberg of the Institute for Advanced Study and two coauthors. They studied situations where there's a background magnetic field filling space. To get the gist of how UV-IR mixing arises here, imagine a pair of oppositely charged particles attached by a spring and flying through space, perpendicular to the magnetic field. As you crank up the field's energy, the charged particles accelerate apart, stretching the spring. In this toy scenario, higher energies correspond to longer distances.

Seiberg and company found that the UV corrections in this situation have peculiar features that illustrate how the reductionist arrow can be spun round, so that the IR affects what happens in the UV. The model isn't realistic, because the real universe doesn't have a magnetic field imposing a background directionality. Still, Craig has been exploring whether anything like it could work as a solution to the hierarchy problem.

Craig, Garcia Garcia, and Seth Koren have also jointly studied how an argument about quantum gravity called the weak gravity conjecture, if true, might impose consistency conditions that naturally require a huge separation between the Higgs mass and the Planck scale.

Dubovsky, at NYU, has mulled over these issues since at least 2013, when it was already clear that supersymmetry particles were very tardy to the LHC party. That year, he and two collaborators discovered a new kind of quantum gravity model that solves the hierarchy problem; in the model, the reductionist arrow points to both the UV and the IR from an intermediate scale. Intriguing as this was, the model worked only in two-dimensional space, and Dubovsky had no clue how to generalize it. He turned to other problems. Then last year, he encountered UV-IR mixing again: He found that a naturalness problem that arises in studies of colliding black holes is resolved by a "hidden" symmetry that links low- and high-frequency deformations of the shape of the black holes.

Like other researchers, Dubovsky doesn't seem to think any of the specific models discovered so far have the obvious makings of a Kuhnian revolution. Some think the whole UV-IR mixing concept lacks promise. "There is currently no sign of a breakdown of EFT," said David E. Kaplan, a theoretical physicist at Johns Hopkins University (no relation to the author of the CKN paper). "I think there is no there there." To convince everyone, the idea will need experimental evidence, but so far, the existing UV-IR mixing models are woefully short on testable predictions; they typically aim to explain why we haven't seen new particles beyond the Standard Model, rather than predicting that we should. But there's always hope of future predictions and discoveries in cosmology, if not from colliders.

Taken together, the new UV-IR mixing models illustrate the myopia of the old paradigm—one based solely on reductionism and effective field theory—and that may be a start.

"Just the fact that you lose reductionism when you go to the Planck scale, so that gravity is anti-reductionist," Dubovsky said, "I think it would be, in some sense, unfortunate if this fact doesn't have deep implications for things which we observe."

DOUGLAS FOX

The Coming Collapse

FROM *Scientific American*

ON DECEMBER 26, 2019, Erin Pettit trudged across a plain of glaring snow and ice, dragging an ice-penetrating radar unit the size of a large suitcase on a red plastic sled behind her. The brittle snow crunched like cornflakes underneath her boots—evidence that it had recently melted and refrozen following a series of warm summer days. Pettit was surveying a part of Antarctica where, until several days before, no other human had ever stepped. A row of red and green nylon flags, flapping in the wind on bamboo poles, extended into the distance, marking a safe route free of hidden, deadly crevasses. The Thwaites Ice Shelf appeared healthy on the surface. But if that were the case, Pettit wouldn't have been there.

Pettit was studying defects within the ice, akin to hidden cracks in an enormous dam, that will determine when the ice shelf might crumble. When it does, the rest of the West Antarctic Ice Sheet behind it could flow right into the ocean, pushing up sea levels around the planet, flooding coastal cities worldwide.

From a distance, the ice shelf looks flat, but as Pettit walked, she saw the guide flags ahead of her rise and fall against the horizon—a sign that she was walking across an undulating surface. To Pettit, a glaciologist at Oregon State University in Corvallis, this was significant. It meant that the ice's underside was a rolling landscape—not what anyone expected. In satellite images, the center of the ice shelf looks stable. But it isn't, Pettit says: "There are five or six different ways this thing could fall apart."

The Thwaites Ice Shelf begins where the massive Thwaites Glacier meets the West Antarctic coast. The shelf is a floating slab of ice, several hundred meters thick, extending roughly 50 kilometers into the Southern Ocean, covering between 800 and 1,000 square kilometers. For the past twenty years, as the planet has warmed, scientists using satellites and aerial surveys have been watching the Thwaites Ice Shelf deteriorate. The decline has caused widespread alarm because experts have long viewed the Thwaites Glacier as the most vulnerable part of the larger West Antarctic Ice Sheet. The ice shelf acts as a dam, slowing its parent glacier's flow into the ocean. If the shelf were to fall apart, the glacier's slide into the sea would greatly accelerate. The Thwaites Glacier itself holds enough ice to raise the global sea level by 65 centimeters (about two feet). The loss of the Thwaites Glacier would in turn destabilize much of the rest of the West Antarctic Ice Sheet, with enough ice to raise sea levels by 3.2 meters—more than 10 feet.

Even the most optimistic greenhouse gas emissions scenarios indicate that by 2050 humanity will likely be locked in to at least two meters of sea-level rise in the coming centuries. That will put the homes of at least ten million people in the US below the high tide line. If the Thwaites Glacier collapses and destabilizes the heart of West Antarctica, then sea-level rise jumps to 5 meters, placing the homes of at least twenty million US people and another fifty million to one hundred million people worldwide below high tide. Although Sacramento, Calif., is not the first city that comes to mind when imagining sea-level rise, it would lose 50 percent of its homes as ocean water pushes 80 kilometers inland through low-lying river deltas. The fate of thousands of coastal towns worldwide hangs on events unfolding in Antarctica right now.

Since 1992 the glacier has hemorrhaged a trillion tons of ice. It is currently losing an additional 75 billion tons of ice every year, and the rate is increasing. What happens next, however, depends on processes that can't be studied from the air—flaws within the shelf that could break it apart, accelerating the glacier's demise. That's why, in 2018, the British National Environmental Research Council and the US National Science Foundation launched a $50 million effort called the International Thwaites Glacier Collaboration to study the glacier and its ice shelf up close.

The collaboration involved eight research teams, including

one that reported this September that the glacier was retreating faster than had been predicted just a few years ago. Two of the teams visited the Thwaites Eastern Ice Shelf between November 2019 and January 2020. Pettit's team examined the central part of the shelf, looking at structural defects and ocean currents underneath. I accompanied her team as an embedded journalist, earning my keep with unskilled labor, much of it involving a snow shovel. Another team investigated the back edge of the ice shelf along the continent's submerged shore, sending a remotely operated submarine down a narrow hole to explore a crucial environment hidden under 600 meters of ice, where the shelf is melting most quickly. The results paint a worrisome picture. The ice shelf "is potentially going to go a lot faster than we expected," Pettit says.

Antarctica's ice sheet has consistently surprised those who study it. In February 1958, researchers in West Antarctica, 700 kilometers inland from the coastline, drilled four meters into the snow, lowered in 450 grams of explosives, and detonated it with a muffled *fuff* that sprayed snow in the air. Geophones sitting facedown on the ice recorded the sound waves that reflected off the hard ground far below. By measuring the return time, Charles Bentley, then a graduate student at Columbia University, made a shocking discovery: the ice in this location was more than 4,000 meters thick—several times thicker than anyone expected—and rested on an old ocean floor 2,500 meters below sea level.

By the 1970s researchers were flying ice-penetrating radar in airplanes that crisscrossed the region. The scattered surveys confirmed that the West Antarctic Ice Sheet sits in a broad basin, deepest toward the center, with large glaciers spilling into the sea through gaps in the basin's outer rim. Even as scientists testified to Congress in the late 1970s about carbon dioxide and the dangers of global warming, most of them didn't think that Antarctica would lose its ice anytime soon. But in 1978 John Mercer, a glaciologist at the Ohio State University, sounded the alarm that West Antarctica represented "a threat of disaster." If the ice sheet lost the shelves separating it from the sea, it might crumble far more quickly than people imagined. Three years later, Terry Hughes, a glaciologist at the University of Maine, called out two specific coastal glaciers—Thwaites and Pine Island—as "the weak underbelly" where the collapse of the ice sheet would most likely begin. A pair of papers published in 1998 and 2001 by Eric Rignot, a glaciologist at NASA's Jet Propulsion Laboratory, showed that these two glaciers were indeed thinning, melting from beneath, allowing ocean water to intrude farther inland under the ice.

Additional aerial surveys since then have shown that the Thwaites Glacier is especially troubling. The ground underneath the glacier is a relentless slope that drops deeper as it moves inland from the outer, seaward edge, allowing warm ocean water to slide under the glacier, melting it from below. As the ice thins, losing weight, it is also expected to lift off the bed and float on the intruding warm, dense water, allowing the water to penetrate even farther—eventually reaching the 2,500-meter trench at the heart of the continent. If that happens, "you're going to unload the West Antarctic Ice Sheet," says Ted Scambos, a glaciologist at the University of Colorado Boulder, who traveled with Pettit's team in 2019–2020.

The glacier flows into the sea in two arms that move at different speeds. The "fast arm" on its western side is a fragile, floating "ice tongue." In satellite images it resembles a shattered windshield, composed of hundreds of icebergs a kilometer or two across drifting into the ocean. The "slow arm," on the glacier's eastern side, is a smaller ice shelf that for years seemed more stable. The front edge butts into a submarine mountain ridge 40 kilometers off the coast. This ridge acts like a doorstop, creating back pressure that holds the ice shelf together.

Pettit and her team chose the mountain-buttressed eastern shelf for their expedition. In satellite images, the shelf's central region appeared relatively stable, its surface smooth enough for small ski-mounted planes to land. A pair of mountaineers could scout for hidden crevasses and establish safe routes, allowing the team to move around freely. Pettit worried that visiting an apparently undamaged part of the ice shelf might limit their opportunity to learn something new. She didn't need to worry.

Antarctic fieldwork requires sending tons of fuel, food, and survival gear ahead of time. The field team has to be supported by layers of transport, workers, and staging camps. All told, the Thwaites research expeditions required several hundred thousand kilograms of equipment and supplies delivered by ships, planes, and convoys of tractors towing sleds across hundreds of kilometers of ice that had been searched ahead of time for crevasses. The British Antarctic Survey and the US Antarctic Program staged some of that gear a year or two in advance. But in Antarctica, even this kind of preparation isn't enough to avoid complications.

In September 2019, two months before I joined Pettit's team as they departed for the frozen continent, they received new satellite images showing two new rifts in the ice shelf. These "daggers" originated where the ice collides with the undersea mountain; the rifts had surged inward toward the coast, to within five kilometers of our planned destination. Expedition leaders worried that one of these rifts could rip through the camp, but the team decided to press forward, with a colleague back home tracking the rifts via satellite. After a series of storms delayed the expedition by a few weeks, we reached the Thwaites Eastern Ice Shelf in mid-December 2019. We assembled a row of tents, protected from the constant easterly winds by walls of snow blocks shoveled and hand-sawed from the landscape, and set up gear for what would be a month of arduous work to come.

The first couple of days were relatively warm. Our boots plunged deeply into the slushy snow, and puddles of meltwater pooled up along the tents. A series of giant ice cliffs, eight kilometers away, were visible to the south. Those upheavals marked the zone where the ice cracked and flexed as it transitioned from a grounded glacier into a floating ice shelf.

As the weather cooled and the snow hardened, Pettit made her first long walks, dragging her radar along preplanned lines. The radar provided two-dimensional profiles of the ice shelf's internal layers, like the slices of a hospital MRI scan. Those first glimpses proved far more interesting than Pettit expected.

Her radar showed that layers in the top 25 meters of the shelf were smooth and mostly flat, but below that they suddenly turned jagged. Pettit speculated that the jagged layers had been part of the ice when it juddered across the rocky coastline bed and started to float seaward, perhaps fifteen years before; they were forever imprinted with the trauma of that transition. The smooth layers represented snow that had fallen on top since then, when the ice was afloat.

More surprising, Pettit found that the shelf's underside—a place that human eyes had never seen—looked strangely ordered, as if it had somehow been sculpted intentionally. The underside was corrugated with a series of trenches that ran perpendicular to the direction of ice flow, like waves offshore from a beach. Each trench was 500 to 700 meters wide and cut as far as 50 meters up into the ice, the height of a twelve-story building. "These things are huge," Pettit told me. Oddest of all, the trench walls weren't smooth, as one might expect of melting ice. They were stair-stepped terraces, with a series of vertical walls each five to eight meters tall, like the sides of an open-pit mine. "We don't know what these stepped things are," she said.

These stair-stepped trenches had escaped detection in previous surveys. Airborne radar measurements are taken from planes moving at least 150 kilometers per hour, so each reading is an average over a long swath of ice. Pettit dragged her radar at a stately three kilometers per hour, allowing her to capture a much finer-grained picture.

As Pettit was getting her first look at the strange terraced structures, her colleagues were starting to see hints of another unexpected observation: the bottom of the ice was not melting the way they expected. On January 2, I wolfed down a breakfast of dehydrated porridge with Christian Wild, a postdoctoral scholar who works with Pettit. He and I then drove a snowmobile out into a frigid snowfall. The sound of the engine was muffled, and the wan light seemed to seep in from all directions, leaving no shadows, no texture, and no hint of the approaching bumps that we trundled over. We steered along our GPS line, with just enough visibility to see each new flag appear silently into view, then dissolve behind us in a gentle slurry of snowflakes.

At a series of stops, Wild used high-precision radar to measure the thickness of the ice shelf, accurate to a few millimeters. He had already measured the same points a week earlier. Because satellite estimates suggested the ice shelf was thinning an average of two or three meters per year, he expected to find the ice three to six centimeters thinner than the week before. To his astonishment, he saw almost no thinning. "It doesn't make any sense," he said toward the end of a long day.

Back at camp, other team members prepared to measure the temperature of the ocean currents flowing under the ice shelf. Over several days they tossed 6,000 kilograms of hard snow, one block at a time, into a canvas-sided tank the size of a large hot tub.

They melted the snow and heated the water, then used it to make a hole as wide as a dinner plate 250 meters down through the shelf. Scambos lowered a string of sensors through this hole into the ocean water below. For the next year or two this sensor station, powered in part by solar panels on a small steel tower, would measure the water temperature, salinity and currents.

Initial readings showed that warm, dense water was indeed flowing under the shelf. At two degrees above freezing, it should be "enough to melt many meters of ice over the course of a year," Scambos said. But the ice wasn't feeling the heat. A layer of cold water sat up against the shelf's underside. Because that water came from the melting of glacial ice (which itself comes from snow), it contained little salt, so it was buoyant, hugging the bottom of the shelf and shielding it from the warmer, saltier water below.

By the end of the expedition Pettit's team had encountered a series of revelations that defied previous views of the ice shelf. First, its underside was eroded with deep trenches, and the slopes of those trenches were organized into stair-stepped terraces. Second, the ice didn't appear to be thinning at the points Wild measured, which disagreed with satellite surveys. Finally, the underside of the shelf didn't seem to be feeling heat from the deep ocean, because it was insulated by a layer of cold, buoyant water. This set of findings was difficult to explain, but another research expedition, operating not far away, would help make sense of the surprises.

Eight kilometers southeast of Pettit's camp, the other group of scientists was getting a first look at the ice shelf's grounding line— the long contour of ground where ice lifts off the land and floats on the sea. In this hidden place, scientists believed the underside of the ice was melting most quickly.

On January 11, 2020, researchers at the camp lowered a black-and-yellow cylindrical vehicle, as wide as two hands and 3.5 meters long, by cable into a narrow hole in the ice. Engineers led by Britney Schmidt, a planetary and polar scientist now at Cornell University (then at the Georgia Institute of Technology), had spent eight years developing this remotely operated vehicle, called Icefin. They had driven it under sea ice more than a meter thick and under the edges of two small ice shelves, where it could be winched out by cable if it got stuck. But they had never lowered this precious object through such a massive slab.

Schmidt sees Icefin as a prototype of a probe that will one day explore vast bodies of water in the outer solar system, hidden underneath 10 or 20 kilometers of ice on Jupiter's and Saturn's moons. In Antarctica, Icefin would measure the ocean temperatures, currents, and rates of melting under the ice. Perhaps more important, its video cameras and sonar would allow the researchers to visually explore this remote environment. Schmidt wasn't looking to validate any of Pettit's observations per se, but the two researchers were working relatively close by, on the same ice shelf, so serendipity could play a role.

After descending through the 600 meters of ice, the vehicle emerged into a layer of ocean water only 50 meters deep. Schmidt, sitting in a nearby tent, steered Icefin with her thumbs on the controller of a PlayStation 4 console. The glassy ceiling of the ice's underside scrolled past on her video monitor as Icefin glided along, sending video up its fiber-optic tether. For eight hours Schmidt guided the vehicle as far as two kilometers from the borehole, into narrow spaces where less than a meter of water separated the ice above from the gravelly, gray-brown seafloor below. This was newly exposed seafloor; the thinning ice had pulled away from it only a few days or weeks before. An occasional fish or shrimp drifted by.

In most places, the currents were sluggish, and close to the ice the water was stratified. As the vehicle approached the grounding line, the water near the ice was at most one degree Celsius above freezing, even though warmer water lay only a few meters away. Icefin's measurements suggested that the underside of the ice was melting at a modest rate of about two meters a year. In some places, meltwater had refrozen onto the bottom of the glacier, revealing a distinct layer of crystal-clear ice, several centimeters thick. Satellite observations had shown this region rapidly thinning, so the findings were at odds with the team's expectations, says Keith Nicholls, an oceanographer at the British Antarctic Survey, who co-led the research at the camp. The overall lack of melting was puzzling, he said: "Extraordinary, really."

As Icefin swam around, it occasionally encountered a clue that would help explain not only these unanticipated observations but also what Pettit's team had found. Cruising slowly along the shelf's fairly flat underside, Icefin came across a vertical wall cut up into the ice—a stair-stepped terrace like Pettit had seen in her radar traces. And the ice on the terrace walls seemed to be melting far more quickly than the surrounding horizontal underside. In the video, there were blurry ripples in the water, where Icefin's spotlight refracted through gushing eddies of saltwater and freshwater swirling together. Icefin also frequently found dark cracks gaping in the ice—basal crevasses as wide as 100 meters. Schmidt steered Icefin up into several of the crevasses, and there again, she found the water swirling and blurry, suggesting the ice may have been melting quickly.

At the December 2021 American Geophysical Union (AGU) meeting in New Orleans, Schmidt's team presented a careful analysis of Icefin's data, confirming that the vertical ice surfaces are playing a pivotal role in the demise of the Thwaites Ice Shelf. Peter Washam, a research scientist at Cornell, reported that the terrace walls were melting five times more quickly than horizontal ice surfaces, losing 10 or more meters of ice a year. The crevasse walls were melting even more quickly—up to 10 times as fast, losing 20 meters of ice a year. Washam noted that the water currents became turbulent as they encountered these steep surfaces, and this brought water into contact with the ice in ways that more efficiently melted it.

The vertical steps may originate from subtle ups and downs present on the ice's underside when it first pulls up from the bed along the grounding line. The ice might fracture and melt more quickly in these uneven spots, steepening the slope—which increases the melt rate, causing the slope to steepen even more, until it forms a terrace wall that is nearly vertical. As ice melts from these vertical surfaces, the terrace walls migrate horizontally, Scambos says. A basal crevasse that is 10 meters across might widen to 30 or even 50 meters within a year. The melting of the Thwaites Ice Shelf's underside isn't a uniform process; it is highly localized, directed by the topography interacting with currents.

If most of the melt is happening on the vertical ice faces, that could help explain why Wild saw no signs of thinning in many of the places he measured. After returning home in 2020, Pettit plotted Wild's points on her radar survey lines showing the terrace walls. In each case, Wild's measurements fell some distance from the closest wall, in a spot where the ice base was horizontal and so maybe not melting much. This isn't unusual, Pettit says, because the walls are spaced far enough apart that Wild was unlikely to hit one by chance. The instrument station that Scambos left behind also seems to be located some distance from the nearest wall; it, too, has shown very little ice thinning.

If the vertical walls are melting quickly, they should also be migrating horizontally across the ice base, Pettit says. At some point, one of those vertical faces will sweep past Scambos's instrument station, "and we will see a huge amount of melt in a short time," she says—perhaps eight meters in a week. "If we see that, it would be supercool."

Schmidt's observations may also explain another feature of the terraced trenches Pettit saw near camp. After Pettit returned home, she examined her radar traces and noticed something peculiar: in the highest segment of a trench, she often saw a stack of inverted U-shaped radar reflections—the classic signature of a crevasse penetrating up into the ceiling. This might occur because the thinner ice over a trench sags like a flimsy bridge; as the ice flexes downward, its bulging belly cracks open. This newly formed basal crevasse may pull in warmer water from below. That would cause the walls of the crevasse to melt and migrate outward, widening until its ceiling is broad enough that it also sags and cracks open—a repeating cycle that could drive cracks ever farther into the ice above.

The massive terraced trenches may have started out as individual basal crevasses, like the ones Schmidt saw eight kilometers upstream at the grounding zone. When Elisabeth Clyne, then a graduate student at Pennsylvania State University, examined radar traces from around the grounding zone, she saw signs that as crevasses moved farther out toward the sea, at roughly 600 meters a year, they were already starting to grow wider and taller through cycles of melting, sagging, and cracking. She reported her analysis at the 2021 AGU meeting in New Orleans. Pettit suspects that these trenches may eventually penetrate all the way up through the shelf or at least cut far enough up through the ice that the shelf becomes prone to breaking from other stresses. This process could splinter the shelf into an unsteady mass of giant, shifting shards that will no longer stabilize one of Antarctica's largest glaciers.

Although Thwaites's western ice tongue lost 80 percent of its area in the past twenty-five years, the eastern shelf shrunk only about 15 percent. Its seaward snout remains pressed against the undersea mountain ridge, which crests roughly 400 meters underneath the ocean's surface. The pressure from this "pinning point" holds the ice together, but the status quo may not last much longer.

In February 2022 Wild published an analysis of satellite measurements showing that the front face of the ice in contact with the underwater mountain ridge is thinning by 30 centimeters a year. At that rate, it will lift off the top of the mountains in the next ten years. Wild expects that when this happens, the eastern ice shelf will rapidly "disaggregate" into a flotilla of icebergs. But it may meet its end even sooner. If focused terrace melting is driving cracks upward through the ice, that could amplify the mechanical stresses that are already tearing at the shelf.

Massive splintering is already happening just upstream of the mountain ridge. Over the past decade the ice there has fragmented into a logjam of long shards held together only by pressure and friction. A series of satellite images, stitched into an animation by Andrew Fleming of the British Antarctic Survey, shows that these shards are sliding past one another with increasing ease. As a result, the splintering shelf is starting to deform and flow around the mountain ridge more quickly and in new directions, like a river that parts as it flows around a boulder. The mountain—once a stabilizing buttress—is now acting as a wedge, sending several "dagger" rifts surging back toward land. These are the same rifts that we saw via satellite just before we left for Antarctica in 2019.

"The thing is falling apart," says Karen Alley, a glaciologist at the University of Manitoba in Winnipeg, who published an analysis of these ice-flow patterns in November 2021. Even if the ice disconnects from the mountain ridge more slowly than expected, another scenario could doom the shelf. Those dagger rifts could keep lengthening until they intersect with the rising trenches advancing seaward from shore. This intersection of structural defects could lead to a shattering of the entire shelf.

In every scenario, the eastern ice shelf will meet a fate similar to the western ice tongue: its constituent shards will disconnect and drift away. Once that happens, the eastern trunk of the Thwaites Glacier will break away from its pinning point, and the western trunk could also speed up. "This whole thing is going to go much faster once the ice [shelf] is all cleared out," Scambos predicts.

*

Pettit's team left Thwaites in late January 2020, but they are still monitoring the shelf's health using solar-powered instruments they lowered into the ocean through holes drilled through the ice. In January 2022, Scambos and Wild returned to our camp site for a few chaotic days to retrieve the data. Antenna and solar towers that once rose seven meters above the ice were mostly buried in hard, icy snow. Scambos, Wild, and two other workers used ice-penetrating radar to find the buried instruments. They then chain-sawed narrow pits six meters down into the ice to retrieve the treasured data cards.

In hopes of getting another year of data out of his instruments, Scambos reinforced the steel towers that had been bent like paper clips and reset the modems that had been fried by static discharge during windstorms. Sensors on the towers had detected winds up to 250 kilometers per hour—nearly Category 5 hurricane speeds and twice what Scambos expected.

GPS units from those stations show that in the two and a half years since they were installed, the ice shelf's seaward movement increased from 620 meters a year to 980 meters a year. As Scambos and Wild gazed down from their Twin Otter plane this past January, they spotted several new tears in the shelf—three kilometers long and several hundred meters wide—where it lifts off the seafloor. Ragged cliffs of ice tilted 50 meters up into the air, exposing deep layers that had not seen daylight for thousands of years. "I think it's losing contact with everything that used to be bracing it," Scambos says. Not only is the ice shelf separating from its pinning point. As it speeds up, it is also stretching and tearing away from the glacier upstream.

The team was so alarmed that Pettit and Wild decided they will return this December to install a new instrument station: "BOB," short for Breakup Observer. They hope BOB will survive long enough to record the final throes of the ice shelf as it fractures into shards. It might not take long.

Scambos speculates that as Pettit and Wild camp on the ice shelf in December, they may wake up one morning to find themselves on a free-floating iceberg. "As long as they're not near one of the rifts, they're not even going to know" at first, he says. Any sounds or vibrations from a crevasse breaching the surface from below might be muffled. Subtle clues will gradually alert them. As the iceberg slowly rotates, their handheld GPS will seem to guide them in the wrong direction, and the sun might also move the wrong way. "You're on this giant white lily pad," Scambos says, "and your only reference is that you're used to having the sun in a certain place at a certain time of day."

My Metamorphosis

FROM *Harper's Bazaar*

FOR YEARS IN my twenties, I wore the wrong shoe size—a 7 when I am actually closer to an 8. I did not know this until I discovered a bunion on my left foot and began ranting to my boyfriend, T, about how all shoes are uncomfortable, how my toes pinched and heels bled, how early hominins may have been right to squelch their naked toes in the soft earth.

T stared at me with polite confusion. "I don't think shoes are supposed to be uncomfortable," they eventually responded. They suggested I try on a pair of their shoes, a size up from what I wore. When I slipped my feet inside the shoe, I felt nothing: no squeeze, no pinch, just space. I walked around our apartment, wobbly until I wasn't. This revelation now seems mortifyingly obvious. But I had become so inured to my ambient discomfort that I assumed everyone felt this way—inconvenienced and constricted by the friction between our bodies and what we use to conceal them.

I have always been disillusioned with the limits of human growth. Even as babies we resemble our future selves, our skin merely stretching, furring, and wrinkling as we age. But insects have a much more fantastic notion of growing up. Their rigid exoskeletons cannot expand with growth, so they molt. They shed old skins and form new ones that are billowing and soft, able to hold more body than before. Some insects like praying mantises undergo transformations with some bodily continuity. They hatch from eggs into tiny wingless adults called nymphs. Nymphs molt into more nymphs, growing larger and sprouting wing buds until the final molt, when the adult emerges like a new, green leaf. But the more famous kind of metamorphosis is more total—that quintessential transmogrification from a caterpillar to chrysalis to butterfly. Born as an egg, growing up as a larva, congealing into a pupa, and unfurling into a winged adult.

It's idyllic to think that all transitions could be this easy: one form sliding cleanly into another, implying a kind of progression. What could be more aspirational than wings? But molting is no easy feat. A transforming insect cannot eat. Sometimes it cannot move. Mayfly larvae must stop breathing for up to an hour as their exoskeletons slip off. The firebrat, a tiny torpedo that feeds on bookbindings, molts as many as sixty times in the few years it lives. Even for insects, there is no one way to transform. These changes can be creeping; some caterpillars already have tiny wings enclosed, unseen within their bodies. Most caterpillars carry no trace of their past selves after molts. Many will eat the skin they shed, swallowing a past silhouette. But some species cannot bear to part with their old selves. In Australia and New Zealand, a moth called the gum-leaf skeletonizer hoards its old heads, stacking them atop its current head like de-nested Matryoshka dolls. These headpieces grow taller with each molt, tapering into a tower of dead heads that can be nearly half as tall as the caterpillar is long. The display is certainly grotesque. But the many-headed tiara may help the caterpillar avoid the jaws of a predator, offering the gift of a more threatening aura and, perhaps, an extra chance of escape.

Maybe this is why I, too, hoard remnants of my past selves. My apartment teems with objects I will never wear again but am afraid to throw away: a sequined chartreuse dress, an unopened vial of hibiscus perfume, a pair of crocodilian heels in a dusty box. Maybe Sabrina—a name that conjures a mold I often do not fit, a name that many of my friends do not use—is also a head I will cast off when I am ready. Maybe hoarding these things makes molting easier, a way of holding on to my past selves. Maybe I will find my way back to them in new ways.

Like all caterpillars, the gum-leaf skeletonizer caterpillar eventually stiffens into a pupa, the protective cocoon in which it will dissolve its body into unrecognizable goo. Here, it reshapes itself into an adult, cells dividing into wings, antennae, legs, and genitals. If you cut open the cocoon, the once-caterpillar would be little more than soup. But if you give it time, something as miraculous as a moth might emerge—silky brown wings corrugated like tree bark.

For a few months now, each night before bed, I smear my skin with an ooze that has the power to reshape my body. At first, it felt like nothing was happening, my human form as rigid as ever. But lately, I can feel myself shifting: thrumming gravel in my voice, pimples on my chin, a pinkish bulge that is now larger than an almond, a newly incandescent desire for sex. But unlike something shielded in a cocoon, my soft body is out in the world. Everyone around me is witness to my transformation, whether they know it or not. I karaoke through my voice cracks, hovering between octaves in search of the safest place to land.

In truth, I do not know where that will be. But I know this: I deserve to be comfortable. I deserve to feel at home in my body and what surrounds it. Because in spite of all the voice cracks and pimples and other embarrassments of a second puberty, I have never felt lighter, more free. Because molting is an act of love. It is a promise that you will grow into yourself and your wild and unexpected future. It is a promise to trust the process, that even mired in goo and guts, something like bliss may wait on the other side.

SARAH GILMAN

The Bird and the Flame

FROM *Audubon*

PEOPLE WHO LOVE Big Basin Redwoods State Park remember it as a refuge. A place cool and damp and dark, crowned with frequent fog and layered branches of redwood, Douglas fir, oak, and madrone. But now there's little shelter to be found here.

On a 90-plus degree day in July 2021, Portia Halbert steers her Prius into the park through a tunnel of dense forest. The reprieve is brief: When we enter the burn zone, it's as if someone has peeled off the roof. The ambient temperature rises, and verdant understory gives way to burnished copper. Halbert parks at a high overlook and leans out the window. From ridgetop to ridgetop, the view is mostly skeletal black trunks.

"How many trees do you see that have greenery in their canopy?" Halbert asks. Maybe twenty, maybe thirty? "There is no way you can look at this and go, 'Everything is hunky-dory.'"

The transformation of the rumpled valley below began the previous summer with a heat wave that struck the Central California coast, where Big Basin encloses an 18,000-acre swath of the Santa Cruz Mountains. Lightning storms rolled in the night of Saturday, August 15, 2020, strafing the state with thousands of dry strikes—and starting twenty-seven fires in and around the mountains.

Halbert had just cut into her bathroom wall for a remodel when she heard the news. At first she thought the flames might do some good. A senior environmental scientist with the Santa Cruz District of California State Parks, Halbert has worked with prescribed burns since her start with the agency in the early 2000s. The professed pyromaniac was thrilled when one of the fires entered an

area of Big Basin she had hoped to burn to improve wildlife habitat, saving her the trouble. It stayed low to the ground, clearing
out brush and accumulated dead branches, trees, and other debris, just as she'd hoped.

Then the winds came. On Tuesday evening several fires tan-
gled into a single shrieking fury that raced through the treetops.
Within twenty-four hours most park infrastructure was gone, including a small town's worth of buildings that has served up to one
million annual visitors, some of whose families had come here for
generations. When the CZU Lightning Complex Fire was finally
contained in late September, it left a scar encompassing 86,500
acres, including 97 percent of Big Basin.

The change boded ill for the four hundred to five hundred
Marbled Murrelets that congregate offshore here each spring
and summer—another of Halbert's responsibilities. The mottled,
robin-size birds spend most of their lives at sea. But when it's time
to nest, the species makes the wildly improbable choice to fly
more than a dozen miles into towering coastal forests from here
to Alaska. They lay a single egg directly on high, wide branches
in ancient conifers, softened with lichen and mosses and hidden
behind a screen of needles. This secretive behavior helps protect
their offspring from predators. It also makes them vulnerable to
logging and development; the species was listed under the federal
Endangered Species Act in in the Lower 48 in 1992.

Marooned far south of the approximately twenty-three thousand other Marbled Murrelets nesting in California, Oregon, and
Washington, the Santa Cruz population has historically depended on Big Basin. The park held the largest concentration of
big, old trees in the Santa Cruz Mountains. But in recent decades,
booming tourism has threatened the bird's survival here. And the
CZU Fire piled on yet more trouble, scorching their remaining
local nesting habitat.

The fire's aftermath was difficult, Halbert says as we descend into
the park's core. "You'd have to brace yourself for it." She points
to stump-covered slopes near the road where workers are in the
process of removing more than twenty-eight thousand dead trees.
The whine of chainsaws fills air that still smells smoky nearly a year
later. Rectangles of earth have been excised where structures once
stood, uprooting every possible contaminant along with the phys-

ical anchors of countless memories. All that remains of the park headquarters, built in 1936, is a stone stairway leading nowhere.

Yet there's hope beyond that painful vacancy. Although many of Big Basin's old Douglas firs died, the vast majority of its namesake redwoods survived. And as staff undertake the arduous process of reimagining park infrastructure, they also have an unprecedented opportunity to change for the better how people recreate amid the sensitive old growth.

They face the kinds of choices that managers across the country will have to make as ballooning visitation and global warming combine to threaten the landscapes and creatures that parks shelter. In that sense, Big Basin has become an early experiment on ecological recovery after a climate-charged disaster. But perhaps, above all, the undertaking will test what a public hungry for connection to wilderness is willing to give up in order to help rare species and rare places survive.

Big Basin is California's oldest operating state park, and the model for the rest. Established in 1902 in response to clear-cutting of redwoods, it became a place where people could commune with the 300-foot-tall giants, and with each other. The park's built footprint grew with its popularity; during the Great Depression, the Civilian Conservation Corps erected a lodge, cabins, an amphitheater, and more structures.

As trees fell outside its boundaries, Big Basin and other protected forests, including neighboring Portola and Butano state parks, became sanctuaries for the murrelet—although people didn't yet know it. The bird was an enigma. Ornithologists had little clue where it nested until loggers in British Columbia began finding chicks and eggs among felled trees in the 1950s and '60s. Then, in 1974, a tree pruner working high in a Douglas fir in Big Basin nearly stepped on a downy chick hunkered on a branch 150 feet off the ground. He had found the first recorded tree nest for the species.

Captivated by the murrelet's story, a federal soil conservationist named Steve Singer began devoting much of his free time to studying the birds and became a local expert and consultant. Working with graduate student Nancy Naslund in the late 1980s, and later with the Santa Cruz Bird Club, he and his wife, Stephanie Singer, spent weekends spying on trees in Big Basin that murrelets might favor, discovering a handful of nests. But while monitoring nests provided insight on murrelet natural history, they were too well hidden, scattered, and high up to study in a practical way. So, in the '90s, state park staff launched audiovisual surveys, sending biologists to watch and listen for murrelets at specific inland study sites over a set time frame. Steve Singer helped lead them.

The findings suggested that Big Basin was the Santa Cruz Mountains' most hopping murrelet breeding ground. The birds were so dense around a study site near park headquarters, called Redwood Meadow, that surveyors sometimes logged as many as three hundred "detections," such as sightings or calls, in a morning. But the counts also revealed that the meadow's murrelets were declining, and by 2005 detections plunged to around eight on average.

It was hard to be sure why, but there was one compelling possibility. Years earlier, Naslund had filmed two Steller's Jays flank a murrelet chick and tear it apart in its nest. "They basically double-teamed him," Singer says. Meanwhile, Common Ravens had been on the rise in the Santa Cruz Mountains since the 1980s. Of the twenty murrelet nests discovered there, predators had raided at least seven, and possibly nine more. Corvids plucked offspring from at least four nests.

When researchers looked more closely, they found Steller's Jays were as many as nine times more abundant near campgrounds in the Santa Cruz Mountains than they were in surrounding forest. Ravens were twenty-eight times more common. It made sense: Campgrounds meant accessible human food. Further studies showed that jays gorging on campers' leavings were healthier than their wilder counterparts and had more babies, most of which dispersed into murrelet habitat. And at Big Basin, some two hundred campsites crowded in and around the murrelets' favored old-growth core, along with picnic areas, parking lots, a museum, and other busy facilities—all near Redwood Meadow.

Changing the park layout would have been politically unpopular and financially difficult. So, in 2012, Portia Halbert started the "Keep It Crumb Clean Campaign." It required campers to watch a video explaining the dangers of unattended food, garbage, and scraps and sign a "Crumb Clean Commitment," backed by fines. Signs and animal-proof food lockers went up at campgrounds, along with dishwashing kiosks and special grates beneath water spigots that kept food bits from accumulating on the ground. To supplement these efforts, park staff also killed some ravens with air rifles. It seemed to be effective: By 2020, corvid numbers in campgrounds had dropped and Big Basin's murrelet numbers were rebounding slightly.

Then the CZU Fire hit. As soon as it was safe, Singer visited some good nesting trees he had mapped before the blaze. Of eighteen suitable redwoods, fifteen had survived and still seemed worthy nest sites. All but seven of twenty-two Douglas firs died. Extrapolating across the burned parts of known murrelet areas, Singer estimated the birds lost 33 percent of their nesting grounds. "It couldn't have occurred at a worse spot in the Santa Cruz Mountains," he says. And for Singer, it begged the question: Would this event push the birds over the edge?

Summer is a sleepless season for people who study birds. The weekly audiovisual survey starts forty-five minutes before sunrise; on July 8, 2021, that's 5:12 a.m. Halbert has been up for two hours when she begins to scan the sky at her site, a dirt lot surrounded by thick forest in Portola Redwoods State Park, just north of Big Basin. The first sounds are the yelling of an American Robin and the laugh track of Acorn Woodpeckers. Then a high, clear note cuts through the marine fog above: *keer*.

"There!" Halbert says. She notes into her tape recorder a murrelet bound inland, as she will all other murrelet activity she witnesses over the next two hours. Surveyors at three other state parks in the Santa Cruz Mountains will do the same.

There is lots to listen for besides *keers*. A groan call that is a cross between a creaking door and a kazoo. The *WHUF-whuf-whuf* of wingbeats under the canopy, a sign of a nest nearby. Soon, murrelets are traversing the gap of sky above our heads in pairs and groups, constellations of flight and song. Social hour, Halbert calls it.

Before the 2021 breeding season, Halbert predicted that murrelets would flock to Portola's unburned forests and abandon severely burned habitat in Big Basin and elsewhere. Indeed, Portola is humming. But Big Basin is humming, too, averaging about fifty murrelet detections per visit, higher than before the fire. In fact, detections at all sites are the second highest in recent years. "I was straight wrong when it came to assessing the likelihood of birds coming to use the habitat," she says.

There is more to be upbeat about. That afternoon Halbert sets up a spotting scope in a ruined Big Basin picnic area. Through its aperture is a branch flexed like a bodybuilder's arm in a still-green Douglas fir, where a sunbeam spotlights a fuzzy, motionless ball tucked against the trunk. A week earlier consulting biologist Alex Rinkert saw a murrelet fly low to this place with fish in its bill. He staked out the location until he was certain: The bird was feeding a chick. Amid the devastation, Rinkert had discovered a nest.

The chick doesn't look like much that first visit. But when we meet Rinkert there the following day, we find it plucking away at its down, wisps of which drift through the scorching air like lost snowflakes. The sleek head that has emerged is capped with gray, its throat cream-colored, the bulgy eyes white-lidded and sleepy. Somehow, in two days, this creature has sculpted itself from lump into bird, preening into smoothness, stretching inert limbs into wings and beating them into life. "It's ready to go," Rinkert whispers, a long solo journey west to the sea.

Watching it, it's easy to feel a sense of rebirth despite the soot that smears our faces and cuffs. Already tall green shoots explode from the bases of charred redwoods. Others are covered with fuzzy green foliage—sprouts spurred by disturbance, a phenomenon that arborists call "poodling." Waxy ceanothus shrubs decorate the ground, along with coyote brush, yerba santa, and other fire-following plants that were scarce before the burn. By a creek where flames have burned away needles, scarified seeds, and opened the canopy with blinding sun, thousands of finger-size redwood seedlings rise from the ground. Even Tim Hyland, a Santa Cruz District environmental scientist who focuses on plants, had never seen that here before; the dense foliage had prevented it. The forest will not be a vaulted green cathedral again for a generation, but it's still very much alive, becoming something new.

Halbert is careful to temper optimism. The nest and monitoring site lie within a nine-acre patch where the fire burned less severely, sparing the canopy. Logically, returning murrelets might concentrate there rather than choose fire-blasted sites with little cover from predators. "My theory is that we had a busy season at Big Basin because most of the murrelet habitat is gone," Halbert says. Singer, meanwhile, believes the birds' abundance indicates their loyalty to their nesting grounds—even torched ones. The fire's true impact may take years to show up in surveys.

But while murrelets are making the most of the destroyed picnic site, other park residents are conspicuously absent. Normally Rinkert would find a handful to a larger flock of jays and ravens hanging out in each campground, despite the Crumb Clean Campaign. Now, as we slowly case for flashes of glossy black feathers, we see mostly the glossy black ripple of burned bark, along with juncos and workers clearing debris. The jays have all but vanished and the ravens have dispersed. Murrelets have a moment of reprieve. Within it, managers are already well into the process of rethinking the park in ways that might help secure the species' future here.

In the year 544 CE, a redwood sprouted in these mountains. When it fell fourteen hundred years later, Big Basin displayed a cross section of its trunk outside the visitor center, a timeline reflecting a redwood's immense lifespan. Tags drilled into packed tree rings commemorated many colonial waypoints: Columbus's and the Pilgrims' arrival in the Americas, the first Catholic mission in San Diego, the discovery of California gold. The monument burned with everything else.

Here, as some historians have noted, there is a literal and metaphorical clean slate in terms of what is built and whose stories guide Big Basin's fate. "It sparks a need to think differently," says Santa Cruz District parks superintendent Chris Spohrer. "We face different problems, not the least of which is climate change, with increased drought and heat and the fires that follow."

Addressing those problems will involve doubling down on efforts to return regular fire to the park. Prairie was once common along this coast, not because of some innate ecology, but because Indigenous nations who lived here maintained it with low-intensity burns, a practice they also used in forests. These carefully managed fires did more than reduce fuel loads, explains Valentin Lopez, chairman of the Amah Mutsun Tribal Band, which has ancestral territories to the south and works to maintain tribal stewardship in the Big Basin region. (The park itself was once home to people who spoke the Awaswas language.) Fires returned sacredness to the land, Lopez says, and nourished meadows that provided forage for deer and elk, material for everything from baskets to houses, and seeds, which were important food.

Spanish Catholic missions forcibly removed local Indigenous people from their homelands around the end of the eighteenth century. Among other injustices, colonizers banned cultural burning, a prohibition that has functionally continued with fire suppression policies under agencies like the US Forest Service and California State Parks in the name of protecting forests and communities. "The park and everyone were just afraid of fire," Lopez says. Over time, there was so much fuel there that "if it got out of control, then it could be horrendous."

But attitudes toward fire began to change, and in the 1970s, Big Basin and the Santa Cruz District introduced prescribed burning within the state park system. The parks have conducted multiple burns since, but Halbert says politics, air quality concerns, and the uncertainty of favorable weather prevented them from burning as often, as hot, or as widely as necessary. Now, though, the CZU Fire has cleaned out the forest enough to make it safe to set larger, more frequent fires. And the destruction of most of the buildings will allow crews to conduct burns in places they couldn't before.

Already the state has earmarked $186 million for rebuilding the park. After an extensive effort to gather public input, staff are considering a shuttle system to move most of the parking out of the sensitive old-growth core, as Yosemite National Park has done to restore water flow in a grove of ancient sequoia. They also envision locating high-impact uses, like corvid-attracting campgrounds and the visitor center, well away from the big, old trees so important to murrelets.

Members of an advisory committee are hopeful that what follows will be bold. The Amah Mutsun, for example, have pushed for tribes to have co-management authority. "Our history has been a difficult one, and the hardest thing in restoring any type of relationship with landowners is trust," says Lopez, a committee member. "Little by little, that trust is building."

Committee member Sara Barth, who leads the Sempervirens Fund, an organization that first helped found Big Basin, notes that park staff have a difficult task ahead of them. "Within the context of a place that is beloved and a place that you expect will burn again, you need to design in a way that is responsive to that reality," she says. "If there is any place in the whole state parks system, and maybe even the country, that could be a beacon for what parks of the future need to be, this would be it."

The real test will come when the park proposes a concrete plan, which must then undergo a formal public review. "It's important that the park continue to communicate that we're making these adjustments to help endangered birds not go extinct," says Shaye Wolf, the climate science director for the nonprofit Center for Biological Diversity. The organization's successful lawsuit against Big Basin over murrelet protections helped secure dedicated funding for the Crumb Clean Campaign and other vital work in 2014. "When people know, they're like, 'That makes sense,'" Wolf says. "'We don't want to contribute to the extinction of the population in these mountains. That's not what we want to do when we camp here.'"

Halbert returns to the nest tree at the edge of twilight. Perhaps a dozen people have gathered in camp chairs with binoculars and scopes, keeping silent vigil for the chick's departure. Rinkert is here, and the Singers, both with colorful knit caps stretched improbably over the hard hats required to sit beneath the fire-scarred trees.

For a long time nothing happens beyond the failing of the light, the worsening of the mosquitoes, the occasional dry crackle of feet shuffling in leaves. Then someone gasps. A silhouette has arrived—a parent with a fish. The transfer takes time. The chick is little more than a white smear. A shift, a rustle, and more gasps. The wiggling smear is gone. I squint through the gloom at Rinkert, who gazes into the sky over my head.

"It dropped off the branch toward us," he says.

"You just lost it in the trees?" Halbert asks.

"It became invisible," Rinkert says, with a slow grin.

Wildlife photographer Frans Lanting pours Prosecco into paper cups and we gather behind his long-lens camera to watch the video he's captured. There it is, crystal clear now: a small bird coming to the edge of all it has known—this green island amid a sea of blackened trees where its parents set its life in motion—stretching its short wings. It shoots straight up into the air and dives diagonally out of frame. Halbert compares it to a hummingbird, a bumblebee. We watch again and again.

There is no way to know whether the murrelet will make it. Such journeys are inherently uncertain, and this one is especially heavy, freighted as it is with meaning for those gathered here. The grim global future. The grief of this burn. The legacy that cut down the bird's kind in the first place. The work that people put in to stem those losses. The ways they've fit their lives to this creature, despite it being unknowable, out of reach, completely other.

I look after the bird's path for a moment, the direction it somehow knew to go. May it reach the ocean. May it find its way through the years ahead, find its way back here, find a way to make a home and a life in this altered world. May we, also.

JOSH MCCOLOUGH

Dislodged

FROM *The Missouri Review*

TWO HOURS SOUTH of Grants Pass, Oregon, we encounter a flashing message board declaring Highway 101 closed. Cars are stopped ahead of us at the top of a hill where the road bends into a dark tunnel of trees near Jedediah Smith Redwoods State Park in Northern California. Two Caltrans officials in hard hats and reflective vests are turning people around. Heavy construction equipment—dump trucks and excavators on flatbed rigs—passes us in the left-hand lane and disappears into the forest.

"This can't be right," I insist, checking my phone. I have not received any alerts. Then again, we just emerged from the mountains, where reception was spotty.

"Can we go around?" my daughter asks.

Google Maps recalculates the quickest alternative route: a three-hundred-mile journey east back through the mountains to the interior of the state, then a return west through the mountains to the coast further south. It estimates the detour to be over seven hours long.

"Nope," I say.

She is a high school student; I am a college English comp instructor. We are in the middle leg of a post-vaccine road trip down the West Coast—Seattle to LA. It is partly a college visit trip for her, partly an excuse to stretch our legs after a year and a half locked down in front of glowing screens. We are from the Midwest and are fed up with the flat, wearying Chicago suburbs—as two-dimensional and enticing as a Zoom classroom. I hate the virus for thousands of reasons, but particularly for what it wrought

on the dynamic experience of a classroom, reducing it to nothing more than glowing foreheads. Postered walls and ceiling fans,
fish tanks, gaming chairs, plaid bedcovers, fairy-lit shelves, rainbow
LED light strips, an occasional bong. But mainly blue LED backlit
stares from deep within a hoodie. Student gazes that go on forever
into a virtual middle distance while you make an utter ass of yourself on camera discussing the elements of a short story or how to
write a literary analysis essay. My daughter and I are on opposing
ends of the same horrific livestreaming scholastic train wreck—she
knows what it looks like to witness a teacher on camera beg, cry,
or yell for someone, anyone to speak up and join in conversation;
I know what it looks like when a young person, already on the
verge—uncertain and unsure—opts out altogether by going dark.

We needed a change of scenery.

"Never underestimate the impact that the physical landscape
has on your mental health," I tell my daughter before our trip,
more as a reminder to me than anything else.

We have just reached the California coast after twisting our way
down the Redwood Highway through rugged, unincorporated
towns—Idlewild, Darlingtonia, Gasquet (pronounced *Gas-kee*).
People in the towns weren't unfriendly, but on the periphery was
a population who, based on the politics spelled out in bumper
stickers on custom trucks, had been living and working remotely
by choice long before the virus. Plenty of handmade "No Trespassing" signs, one of which read "No Trespassing! Iraq War Vet
with PTSD," a dripping AR-15 stenciled in spray paint underneath.
We snaked through sunlit mountain passes along dried-up creek
beds, until a blanket of coastal fog swept over a crest and envel-
oped the highway for a few sudden low-visibility-on-sheer-cliffside
moments. When we emerged, the Pacific opened up before us,
gray and soupy. Fog and cloud cover melded together, giving everything a vaporous edge. Monolithic sea stacks peppered the base
of dark green marine terraces. It was a revelatory moment for us
midwestern pilgrims, who, though we might not have set out on
foot from Missouri, felt an undeniable rush in reaching the end
of the westward road. We rolled down the windows and inhaled
deeply. Then we rolled to a stop at the flashing sign and the line
of cars and people turning back.

We approach the Caltrans official, who repeats the message on
the sign. "Road's closed," she says and hands us a flyer. "You can go

back to Crescent City, or you can proceed ahead and wait in line until the road opens again at one o'clock."

"Okay, but what's happening?" I ask.

"There are active landslides at Last Chance Grade, and crews are working to shore up the highway," she says. "If you want to wait in line, they're distributing bottles of water and granola bars. But once you get in line, there's no turning back."

So be it.

After waiting nearly two years to go anywhere, sitting in our car for a few hours in the forest does not feel like such an imposition. Sometimes in order to move forward, you have to stay put for a bit—one of the many lessons imparted to us from the virus. We turn around and claim our moment at Crescent Beach, where we dip our toes in the frigid Pacific and watch a solitary wet-suited surfer bobbing in the waves. We stock up on water and snacks and head into the forest to wait. A flashing police cruiser escorts a line of cars into the forest at a pleasant minimum speed. It feels as though we're on a guided tour of the redwood forest. We roll down the windows and poke our heads out to look up at the trees. After a few miles, we reach the line of those who ventured before us and stop along the side of the zigzagging road. We don't know how close or far we are from the construction—from Last Chance Grade.

We emerge from the car, and our eyes are directed skyward. On either side of the road are colonnades of redwoods. Above us, cathedrals soar hundreds of feet and block out all but slivers of the gray-fogged sky. I reach for my phone to FaceTime my wife, but there is no reception. Not one bar.

"Is this place for real?" my daughter asks, not for the first time on this trip.

The road cuts an unnatural, gray-paved path through the woods. The coastal fog has followed us into the forest. The tops of the redwoods sway, yet there is no breeze at ground level. It feels like we are underwater. Voices are small but distinct. Clear. One man tells his kids to put down their damn phones for a second and come out and look around. The kids stay in the car. Another man opens his car door, grabs his camera, and aims his lens upward, the camera's shutter rapid-fire clicks. A woman worries about having to go to the bathroom. She wonders if she can hold it until the road re-opens. The man she's with directs her into the forest, and she tells him she'd probably get poison oak all over her privates. Another woman climbs atop her camper and peers into the forest through binoculars in a way that signals she knows what she's looking for. A man emerges from a RV in full spandex; he unhooks a bicycle from the rear rack, straps on a helmet, and turns on flashing LED lights and pedals ahead. "May as well log a few miles while we're waiting," he says as he passes us. Another man opens the door to his SUV, setting free two barefoot toddlers, who wobble onto the road. The man is also barefoot. He lights a cigarette. Someone nearby is smoking pot; this seems as good a place as any to do so. A small group of teens in pajama pants and hoodies walks up the side of the road, happy to get away from their parents.

Right now, the road is connecting us differently than when we drove it. Not ten minutes before, the man in the car behind us tailgated me and honked at me for driving slowly, though we were being paced by a police car. I could see his darkened figure in my rearview mirror throw up both of his hands in a "What the hell?" gesture. Now he gets out of his car, smiles, and says to me, "Not a bad place to be stuck, is it? Just beautiful."

Wanderers, all of us, forced to be still for a bit. To see what is around us and see one another. These are the kinds of friendships forged among strangers in a church parking lot.

The ground on either side of the road is covered greenly in sword fern and redwood sorrel, bracken fern, wild ginger, trillium, and moss. Shoots of yellow monkeyflower rise above the brush cover. Tanoaks and Pacific rhododendrons (a woman—clearly local—from the car in front of us tells us that we missed them in bloom by about a month) grow between the colossal redwoods. They are what we midwesterners might think of as good-sized trees—tall but climbable. Though at the feet of behemoths, they appear wispy and decorative. My daughter and I walk across to the other side of the road and look down upon a ravine. The forest floor is brick red, carpeted with dead, needlelike redwood leaves. The trees creak softly.

Then, a whistle—flat, off-key—breaks through the forest, and another whistle calls back. It sounds metallic. It is constant, like a referee's whistle, but there is no rise time—it starts and ends at full whistle. The whistling surrounds us like the forest itself. Everywhere I turn, it sounds like it's coming from behind me. A long, off-key whistle. Another that calls back.

An oncoming dump truck blows its horn, echoing like an alpine horn through the forest, and people on the road alert one another. Parents gather kids in their arms, and the truck barrels by us in a *whoosh* toward what must be Last Chance Grade.

"Good lord," the woman from the car ahead of us says. "What's his hurry?"

Everything is short on this trip. Tempers are short. Hotels and restaurants and gas stations are short-staffed, short on menu items, short on services offered. Operating hours of restaurants, cafés, and bars are cut short. Grocery stores are short on items. Trucking companies are short drivers. The window of opportunity to move safely about the country is shortening. (The Delta variant is just beginning to spread in the US.) Expectations of a return to absolute freedom are cut short—some states aren't yet open for business; others never closed.

Still, all routes on our West Coast trip are flush with families packed into trucks, campers, cars, and RVs. Luggage racks, boats in tow, American flags frayed and flapping at speed down every road. It almost resembles what "normal" looked like before, until you're reminded how far we have to go still. My daughter and I stop at a diner for lunch. The lights are off, but handmade signs insist "We ARE Open." One of two servers on staff tells us to "sit wherever," so we find an open table. The place is packed. Our server stops to take our order and explains, "Sorry, it's just the two of us. And one cook." The lights are off to save electricity (the owners are clearly short on funds to pay the bills each month during the pandemic). Despite all odds, the server is kind and smiling. She briefly mentions being happy to work again. We don't understand why. The patrons are short on time and patience. Short on tact. *Where's my goddamn cheeseburger? I ordered it like an hour ago. You want me to go back there and make it myself?*

We all fall short sometimes, despite our best efforts.

The whistling in the forest continues. Long and flat. Odd and off-key. Another whistle calls back. I wonder if it might be hikers signaling to one another. My daughter walks along the side of the road, just looking.

I hold the flyer about work on Last Chance Grade and am stuck on the name. Any chances are hard enough to come by these days,

I think. And everything these past couple of years has felt like a last chance. Just leaving my house to scrounge picked-over store shelves for toilet paper felt like a kind of last-chance endeavor. And truly, I am tired of thinking about last chances. What if the last time I saw my parents was my last chance to have seen them? What if the last time I stepped foot in the classroom was the last chance I had to do so? What about that last time I went to a concert and screamed in revelatory joy? Or the last time I sat inside a coffee shop? Or the last time I went anywhere without a mask? The last time I saw my students in the classroom, in spring of 2020, I told them that we might have a week or two of online classes, then would be back in the classroom for the end of the semester. That was right before spring break. My parting in-person words were, "Have a great spring break—see you back here in a couple of weeks!" Now I would really like to have had a chance to say, "I care about all of you; please be safe. Stay with your families or check in on them as much as you can. Love them." We were not given any last chances to do these things until, suddenly, we had no chances for a while.

The whistling cuts through the forest. Over and over again.

The woman from the car ahead of us says, "Ooh, look, banana slugs! They're all over the place."

We haven't noticed them—tiny ground creatures in a mammoth forest—but once we do, it is difficult not to spot them everywhere. Bright yellow or mustard brown, the uncanny (and unfortunate) shape and size of a larger dog's penis, but with eyestalks. They creep about on the ground over dead leaves and hang precariously on low-lying brush like obscene, slimy ornaments. They consume the dead, and in their wake is a trail of slime-nutrients that fertilizes the soil. I crouch down to get a picture of one that is the color of a ripe yellow pepper and see an even bigger one right next to my foot.

I realize that I nearly stepped on it.

I am not a geologist, though I am broadly curious about the reasons why it might not be safe to tread upon parts of the earth, whether it be to preserve the privacy of a wounded veteran or because the ground might give way and wash you into the ocean without warning. Not that we humans are great at heeding warning signs given up by the earth. We exist upon massive lithospheric rafts that float on a layer of plasticine rock. The earth's crust is but the skin of a grape, relative to the rest of the planet beneath us. We are reminded of this when islands burst forth in the middle of the ocean; when a long-dormant volcano awakens; or when World Series games are interrupted by two plates going bump in the night; or when a tsunami arrives, uninvited, to a tropical holiday. These events are unfortunate reminders of precisely who—or what—is in charge here. Still, we too often move through life not considering our size and stature relative to forces and objects that humble us. Geologic time. Plate tectonics. A virus. A couple of degrees' difference in the oceans' temperatures. More rain and less snow. No snow and too much rain. Fire tornadoes. A couple of inches more of the ocean and a few hundred thousand more people underwater.

I tell my daughter, "Stand next to that tree and spread your arms out so we can get a sense of scale."

Some redwoods are hollowed out so cars can drive through them. Not far from where we are is a famous redwood playland (complete with a talking Paul Bunyan) that will cost admission to explore. We don't consider how long it took for this tree to grow so large, but who isn't tempted by a priceless photo or social media op? Our inability to see ourselves as tiny points on a much longer ecological or geological spectrum is our uniquely human blind spot. It's where and how we fall short.

This is what will kill us all, I think, as I click pictures of my tiny daughter at the base of a two-hundred-year-old tree. If last chances are the fuel for redemption, our tank feels so close to empty.

Whistling again.

I long to understand why my daughter and I are stuck in a whistling forest. Why our West Coast road trip itinerary—Leg 4, Day 7—was blown to hell by an ominously named piece of land. What I learn, long after we return home, makes me thankful that I did not know about Last Chance Grade while we were there. A 2015 engineering feasibility study characterizes this stretch of Highway 101 as failing frequently and the ground beneath the road as unstable. To a midwesterner, driving along the edge of the California coast is a vertigo-inducing, heart-palpitating experience anyway. If you are the driver, the fear of falling into the ocean is more omnipresent than the image you had in your head about a fun, carefree, top-down thrill ride along a classic stretch of Americana. If you are a really specific kind of midwesterner, you may obsessively recall grainy dashboard camera videos of cars jettisoning off the Pacific Coastal Highway into the ocean below. No guardrails, nothing stopping the car's launch. Each time the road hairpins and the land slips away and the height above the ocean becomes clear, I get dizzy, while attempting to maintain calm for my daughter, who is in the back seat, also sick. As I recall that drive now, my palms are sweating.

But here lies Last Chance Grade, existing at the intersection of physical and human geography. There have been hundreds of landslides in this area, dating back to the late 1800s. Some of the more recent landslides have been caught on camera and are shocking in their force—their ability in moments to wash away human-made structures engineered to be permanent and unmovable. This three-mile stretch of the 101 undulates, fractures, dips, and ultimately fails because it is built upon four deep-seated landslides that are actively in motion. The highway fails because the ground beneath, part of a large subduction zone, is not stable enough to support a highway. The geography of much of populated California is like this, though, and that a major highway runs across an active landslide may be surprising only to pragmatic midwesterners, who think, "Kind of a silly place to put a road, isn't it?" But that thinking runs counter to the ethos of California, which my daughter and I learn later as we walk around San Francisco and a magnitude 6.0 earthquake hits at the California-Nevada border, causing rock and boulder slides along another major highway while we traipse up Lombard Street and take pictures. We don't even feel it because we aren't standing still.

To the east of the road where we stand is a UNESCO-protected World Heritage Site, home to thousands of animal species in addition to the old-growth redwoods that have existed for up to a couple of thousand years. To the west of the road, a mile or so, is the Pacific. It pounds the base of the cliff upon which the highway has been built, accepting residual detritus from the landslides. This is the physical geography.

Also to the east of the highway—beyond the UNESCO-protected forest—are multiple tribes of indigenous people who have inhabited the land for centuries. The 101 itself is the main artery that supplies communities up and down the coast with food and other essential goods. Block the artery, and food deserts are created. All human inhabitants are taxpayers. All human inhabitants are affected when the road shuts down and will be affected if the road has to be moved. This is the human geography.

The problem of the road has brought together experts in both human and physical geography to consider solutions. After years of economic impact studies, risk assessments, geotechnical investigations, ground surveys, botanical studies, wetland delineations, traffic studies, biological assessments, the road still fails. The ground is still unstable. People, communities, are still left stranded. Doing nothing is not a viable option. Though perhaps if communities are engaged in coming up with a solution together, the devil's bargain will be less difficult to swallow: Cut into some of the most beautiful, ancient, protected lands to move the highway farther east; or tunnel beneath some of the most beautiful, ancient, protected lands to move the highway underground.

The two-mile tunnel is scheduled to open in 2038. As of today, it is estimated to cost $1.3 billion.

The metallic whistling in the forest sounds urgent—a bit like a call for help. I listen for voices—for people calling out—but don't hear anything. I don't know what it communicates. I think it sounds lonely, and then it sounds deeply melancholy. I think it sounds like a warning, and then it sounds like an urgent call for help. Dump trucks speed past us in the opposite lanes and blow their horns; the sounds ricochet off of the trees, reverberating bass throughout the forest. Could the whistling be nothing more than construction sounds ahead of us on Last Chance Grade? I am reminded of a story I heard once on NPR about a scientist in search of the quietest place on earth, free of human-made noise—aircraft, traffic, cell phones, construction, voices. You have to travel so far to get away from human noise. I consider how easy it is to hear other travelers' conversations. People think of forests as quiet places, but they are acoustic marvels. Communication travels efficiently, by evolutionary design. Animal calls seeking a partner in the springtime. Calls warning of predators in the area. Whistling perhaps designed to baffle stranded travelers. I imagine someone up in a tree, blowing a whistle and peering down at me through binoculars, laughing as I turn around to try to find the source.

I remember a story from college of a woman named Julia "Butterfly" Hill, who took up residence in the canopy of an old-growth coastal redwood. Later, I learned that the tree is still there—located a few hundred miles from where we were. She lived in the tree for 738 days on a six-foot-by-six-foot platform to protest a lumber company's clear-cutting practices. In fact, the company's overlogging resulted in a catastrophic landslide that buried much of the town of Stafford in Humboldt County in 1996. She was regarded by the public as a nuisance, an eco-warrior, a curiosity, a crackpot, a neo-hippy, a savior. I remember this. From her tiny platform, she took media calls, debated CNN anchors, responded to mail she'd received from critics and supporters, studied field guides to identify the birds that inhabited trees around her; she let the tree sap cover her feet so that she had better grip while climbing. Loggers shouted vile insults up to her. It was all very loud at the time—everyone had an opinion about her, about the loggers and logging company, about the environment and "environmentalists," who tended to be cast as a fringe, neo-cultist movement. So West Coast.

But since Julia "Butterfly" Hill's tree residency, it has been proven that trees communicate with one another via an underground network of fungi. They work together to survive by transferring nutrients—carbon, nitrogen, phosphorus, hormones, water—to one another. Within a community of trees, there are hubs—mother trees—that nurture their young by way of hundreds of kilometers of fungi below ground. They send excess carbon to the younger seedlings, and if a mother tree is injured or dying, they can send messages to their seedlings to help strengthen them and defend themselves from future issues. Mother trees are vulnerable, though. You take out a mother tree, the system beneath it likely will collapse.

The whistling continues, bouncing off trees.

The distinct whistle remains lodged in my head long after we return from our trip. After the trembling San Francisco, across the interior, seething San Joaquin Valley, down through LA. The whistling follows me. It is a call back to that place—to those hours spent in pause, waiting, looking. But I do not know how to discover the source. I sit with my laptop and some wine and fumble about with far-too-literal search terms.

Whistling in redwood forest.

Whistling noises Pacific northwest redwoods.

Odd metallic whistling redwoods Pacific coast.

Eventually, I find the right combination of words and discover a thread in a forum where others are searching for the same thing. Same location—Jedediah Smith redwoods, Del Norte County, California. Original posters describe the noise as a "referee's whistle" or "a long, electrical whistling" with another slightly off-key callback. I'm excited by this—others heard the same thing. Crowd-sourced responses mean well, sometimes. It is, they say, the trees rubbing against one another. Elk in heat. Bigfoot. Deer. Deer in heat. An owl. Military exercises. Bats. Forestry workers. Mountain lions. A waxwing bird.

A bird. A bird seems like a promising lead, so I search for birds common to that area and become suddenly grateful to the massive online community of ornithological enthusiasts' meticulous dedication to recording sounds. I listen to dozens of bird sounds with my eyes closed. Pacific wren. Acorn woodpecker. Townsend's warbler.

Then I hear the unmistakable, indelible off-key whistling and the callback.

Ixoreus naevius. The varied thrush.

I am overjoyed. I call my daughter out of her room, and declare, "I found it!" I play the sound for her, and she says, "Cool," and recedes back into her iPhone. For me, though, it is a transportive sound. I am back in the forest—in those hours when we were forced to take a good look and listen to where we were. I look up information on the varied thrush, and find it is an ordinary, robin-sized bird. Mostly black with bands of pumpkin orange on its breast, wings, and head. It exists primarily in the Pacific Northwest, though it migrates seasonally up and down the coast when breeding. Still, it is a predominant fixture of the damp green forests along the Pacific, and like grunge, its haunting call is something of a signature sound of the region. It is also held in mythical regard by both amateur and career bird lovers alike. A post by the US Fish and Wildlife Service about the varied thrush quotes ornithologist and illustrator Louis Agassiz Fuertes, who described the varied thrush as "perfectly the voice of the cool, dark, peaceful solitude which the bird chooses for its home as could be imagined." In his 1909 book *The Birds of Washington,* ornithologist William Leon Dawson described the song of the varied thrush as "a single long-drawn note of brooding melancholy and exalted beauty—a voice stranger than the sound of any instrument, a waif echo stranding on the shores of time."

I am entranced by the descriptions of the sound itself. I stack field guides on my table at the library, and I thumb through all of their descriptions of the song of the varied thrush:

> "Song utterly bizarre: long, vibrant, metallic, breathy notes spaced far apart: zeeeeeeng. . . . Zoiiiiiiiiing . . . zeeeerng . . ." (*Smithsonian Field Guide to the Birds of North America*)

> "Song a long, eerie, quavering, whistled note, followed, after a pause, by one on a lower or higher pitch. Call a quivering low-pitched zzzzew or zzzeee and a liquid chup." (*Peterson Field Guide to Birds of North America*)

> "Call a short, low, dry chup very similar to Hermit Thrush but harder; also a hard, high gipf and a soft, short tiup." (*The Sibley Guide to Birds of North America*)

It is an elusive, solitary bird, not easily spotted. By all accounts, the varied thrush likes it that way. How grunge. I stare at pictures of the varied thrush, and it sparks another memory. I recognize the bird somehow, and I can't figure out from what. Eventually, the internet tells me that it is the bird that appears for a few seconds in the opening credits of the '90s television show *Twin Peaks,* which is so fitting, I decide its use must have been on purpose. The varied thrush is the ultimate Gen-X bird.

In the end, it is one p.m., and miles ahead of my daughter and me, blockades open. All down the line, people return to their vehicles. The timeout has ended. I do not want to leave this place, though I want to see Last Chance Grade, maybe to thank it. This diversion will become a centerpiece memory of the trip itself. My daughter and I will recount how we stumbled into a magical interruption on our trip down the coast.

In the end, the line of cars moves forward, and we are pulled along with them. We all move on. We come out of the trees. Out of the banana slug forest. Away from the call of the varied thrush. The road twists and dips through the redwoods until the trees open up to a clearing, and we can finally see it.

In the end, there is a scarred hillside that refuses to stay put, and then a cliff over which things have been falling for many years. Covered wagons, boulders, sediment, stones, cars, trees, dead leaves, mud, construction equipment, banana slugs, fallen redwoods, roots, mycelium. It all slides down into the Pacific. In the end, Last Chance Grade turns out to be neither a place—a pin on Google Maps—nor a natural sight to behold. It is a geological riddle. As the road crosses the grade, we can see car-sized boulders and mounds of soil that have spilled onto it from a recent slide. The road itself becomes nothing more than jagged pavement and compacted dirt—a callback to its original trail state. Above the road, Caltrans pickups and dump trucks and earthmovers and graders and men in hard hats are crawling about the hillside like ants. Thousands of pounds of machinery look barely attached to the earth it seeks to shore up, and I am struck with the familiar sensation of vertigo. In the end, we pass safely across Last Chance Grade—that point of convergence between human and physical geography—a precarious road clinging, like the rest of us, for dear life against all natural forces acting upon it. A waif echo stranding on the shores of time.

VANESSA GREGORY

Bright Flight

FROM *Harper's Magazine*

IT WAS A mild May evening in South Carolina's Congaree National Park, and Raphael Sarfati stood on a trail beneath a loblolly pine he'd chosen as a landmark. The horizon to his left glimmered with light from the nearby city of Columbia, while the tree marked the border of a dark and wild wood. Slanted rays of twilight burnished the trunk, which stretched for perhaps eighty feet before dissolving into an oval of green. The pine, with its fire-blackened base, was probably not yet fifty years old, a juvenile in a bottomlands ecosystem with thousand-year-old bald cypresses. It was the sort of tree Sarfati could recognize in the night, and so it served now as a cue to plunge into the undergrowth toward his field site.

Sarfati is in his early thirties, with curly hair, a compact frame, and a congenial French accent. He moved deftly despite being weighed down by camera bags, tripods, and a bulky poster that resembled a black-and-white checkerboard. Congaree occupies a natural floodplain, and the ground was soft and black and uneven. He stepped high to clear logs lined with turkey tail mushrooms, and to avoid the thick, looping vines that cascaded from the branches above into cushions of dried leaves. When he stopped at a clearing tufted with wispy grass, the woods felt deep and enveloping, though we couldn't have traveled much more than a hundred yards.

"I try to set the cameras up in the same place every night," Sarfati said, dropping his gear in a stagelike space framed by two downed trees. What he wanted to capture, what he'd been waiting to see for the past four nights, was the peak activity of *Photuris fron-*

talis, a species of firefly that ornaments these woods with rhythmic, synchronous light for a few brief weeks each spring.

Firefly synchrony, in which male fireflies congregate and flash together in an astounding mating ritual, was for many years believed to be rare and largely confined to stable populations of the Southeast Asian genus *Pteroptyx.* It was only in the 1990s that the Tennessee naturalist Lynn Frierson Faust introduced biologists to years of her meticulous observations, revealing synchrony to be far more widespread among fireflies of the Eastern United States. Ever since, firefly synchrony has become an increasingly well-known and celebrated spectacle, drawing roughly thirteen thousand visitors to Congaree National Park each spring, and a similar number to the Great Smoky Mountains National Park, where another species, *Photinus carolinus,* glows in a similar fashion.

More than two thousand species of fireflies exist, populating every continent except Antarctica. Their adult lives are brief, often lasting just a few weeks, and their flashes are part of what Faust described as a "desperate mating effort." But in addition to attracting mates, their distinctive flash patterns serve to broadcast distress and even to deceive, as in the case of female *Photuris,* which imitate the flashes of other species' females to lure males and devour them. Male *Photuris* have also been observed to mimic other males, an enigmatic behavior that may represent an attempt to trick the females they know are hunting into mating instead.

Unlike Faust, Sarfati is not a naturalist. He's not a biologist either. He grew up in Paris, the son of an Algerian-born father and the grandchild of Hungarian émigrés, with an early awareness that he was expected to achieve. He grew fascinated by atoms at around age five, and went on to earn a doctorate in physics at Yale, where he studied soft matter (a topic esoteric enough that Sarfati seemed mildly embarrassed whenever I asked about it). During his first postdoctoral fellowship, at the University of Colorado, he focused on polystyrene beads that measure one thousandth of a millimeter. Then he heard about a professor named Orit Peleg, who used physics, computer science, and chemistry to study bees, and transferred to her lab. Last spring marked Sarfati's second year doing experimental field work in Congaree as Peleg's postdoc, having traded his studies of microscopic particles for the realm of living, breathing animals, or what Sarfati likes to call "thinking objects."

His first task in Congaree was to set up cameras to function as "digital eyes," which could sense light as well as humans but perceive depth much more capably. Sarfati wore a long-sleeved shirt from an ultramarathon he'd run in the Rockies and a headlamp around his neck like a pendant. He couldn't complain about the conditions: no yellow flies yet, and only moderate heat. Likewise, he hadn't heard or seen any of the huge wild pigs that had snuffled at the margins of his site the previous year.

He began fiddling with two cameras, positioning one on a tripod beside an American holly and the other, roughly parallel, next to a pine. Then he walked to the tip of an imaginary triangle in front of the cameras, where he placed the checkerboard atop a collapsible easel. He used the board to calibrate the cameras, then started filming and waited for the gloaming to sink into night.

The first fireflies appeared haltingly: just a few random flashes scattered across the understory, like flames from brittle white sparklers. "This is a *frontalis* right there," Sarfati said, pointing at an insect blinking a foot above the ground. "It's flashing about three times every two seconds, like this kind of snappy flash. And it's moving more or less in a straight line. Not exactly, but it's not doing complicated loops or patterns."

Sarfati spotted a trio of *frontalis* drifting past low bushes and made finger guns at them, popping his wrists up and down to match their beat. "Look at these," he said. "One, two, three. Three of them in sync." I mentioned that I'd thought you needed hundreds, if not thousands, of fireflies to trigger synchronicity. "That's what a lot of people think, the peer pressure, that you need a lot of common influences," Sarfati said. "But no, they really want to do it!" Two fireflies, for instance, might be in sync with each other, but not the entire swarm. The famous mass synchronous displays, which tourists in some places must enter lotteries to witness, begin in small, barely discernible groupings. "But if you start to have higher densities, you're connecting all these dots together, these networks of connection and then 'Boom!'"

The sky had grown dark, and a thundering chorus of crickets or frogs arose where there had previously been trilling wrens. At almost nine o'clock, the woods exploded in rhythm, just as Sarfati had predicted. Countless cold points of light flashed and extinguished with uncanny regularity. The synchronicity was so precise that the fireflies resembled a single luminous organism hovering above the forest floor. "Oh, wow," Sarfati said softly. "This is good. It's really good." Puddles of ethereal moonlight had pooled near Sarfati's feet, but they seemed suddenly dull and ordinary. Thousands of *frontalis* glowed and dimmed in every direction, their pulse steady and ancient.

The firefly pioneers of the twentieth century were Elisabeth and John Buck, a married couple who began methodically studying fireflies and theorizing about synchronicity in the 1930s. As a doctoral student, John outfitted a darkroom to simulate a natural cycle of day and night, filled it with fireflies, and set up a sleeping bag so that he could observe their behavior over twenty-four-hour periods. Later, the couple's projects would take them to Jamaica, Borneo, Thailand, and Papua New Guinea. Over their long and celebrated career, the pair studied various types of bioluminescence, including those found in marine organisms, but they always returned to the puzzles of fireflies and synchronicity. "Just what capability communal synchrony demands is a question that does not come easily to us as human beings," the Bucks wrote in 1976, "because we accept our own ability to dance or to march in unison as being second nature."

Indeed, although Western travelers had been documenting sightings of synchronous fireflies for centuries, researchers who preceded the Bucks were often skeptical or even hostile to the notion that invertebrates with millimeter-size brains could engage in such sophisticated behavior. They dismissed reports of firefly synchrony as accidental, as a by-product of humidity or tree sap, or as an illusion caused by a twitchy eyelid.

It's now obvious that firefly synchrony not only exists but has many analogues in nature, where seemingly simple animals organize themselves in staggeringly complex ways. Spiny lobsters form single-file lines on the ocean floor and, like cyclists drafting in a pace line, move faster than any single crustacean might on its own. Vast schools of fish execute intricate high-speed maneuvers— splitting and re-forming behind a predator, for example, or bursting radially—and rarely bump into one another. Giant honeybees nesting on open combs repel predatory wasps by forming spiral waves that spread through the swarm within milliseconds. And most of us are familiar with the sight of starlings wheeling in great dark flocks across the sky, twisting and turning at thirty miles per hour, maintaining a tight choreography even as their numbers swell to the tens of thousands.

How does order arise in the absence of leaders, amid the chaos of thousands of moving, sensing, thinking beings? What, if anything, are these groups trying to achieve? Sir John Bowring, an English governor of Hong Kong, wrote in 1857 that the fireflies' "light blazes and is extinguished by a common sympathy," framing synchronous light displays in plainly human terms. A naturalist writing a half century later proposed that the virtuosity of starlings might be explained by avian telepathy.

Such questions might seem like natural territory for biologists, but they were first embraced by chemists and physicists. They wondered whether mathematical models that explained matter could be adapted to different systems. If biologists traditionally looked at animals as organisms within ecosystems, physicists analyzed them as particles in motion that interacted with forces, modeling the behavior of flocks and schools by using the rules that explain magnets, self-propelled particles, and maximum entropy.

Many of these quantitative researchers were inspired by the dawning science of complexity, a field embodied by the interdisciplinary Santa Fe Institute, which opened in 1984. Santa Fe acolytes chased the notion of universal phenomena that share the same general mathematical descriptions. The idea was that dazzling complexity—be it a chemical reaction or termites building a mound—might arise when individuals in a system follow similar sets of rules at the local level.

A shortcoming of this early work was that it was almost entirely computational and theoretical. Researchers who ventured into the real world to conduct experiments struggled with the limits of what their cameras and computers could do. One famous study that involved filming starlings, by the Italian physicists Andrea Cavagna and Irene Giardina, relied on an image-matching algorithm that took two years to build. Other early researchers did statistics by hand: a staggering proposition when faced with thousands of animals.

"In the lab, I have what are relatively inexpensive cameras right now that are imaging at one hundred frames per second," said Nicholas Ouellette, a professor of civil and environmental engineering at Stanford who studies midge swarms and the flocking behavior of jackdaws. "I can, you know, in a day or two, generate tens of millions of data points that I can do statistics on."

This data revolution lets researchers test the collective behavior models from previous decades and propose new ones. In many ways, the study of animals using quantitative tools and methods is still in its infancy, wide open for discoveries that augment our understanding of evolution, animal communication, animal sensing, and possibly cognition. Many of these insights will start with fundamental questions about natural phenomena that are at once deeply familiar and stubbornly hard to explain. "What is it about a flock that makes it a flock?" Ouellette asked. "What is it about a swarm that makes it a swarm, and not just a bunch of things flying around?"

One evening in Congaree, I watched Sarfati's adviser, Peleg, lift a petri dish containing a male firefly that had fallen on its back. It had switched from its normal, unhurried flashing to a more alarmed frequency, its lantern glowing quickly and urgently, a message of distress as plain as any warning beacon of human design. "In the case of fireflies, the communication is a light pattern, kind of like a flashlight that turns on and off like Morse code," Peleg said.

She was working at a site about a quarter mile from Sarfati, on a gravel track that shone white beneath a nearly full moon. A serene first-year graduate student named Owen Martin had erected two tents. In one, they planned to place individual fireflies to see if they would match their flash frequencies to that of an LED robot; in the other, they planned to introduce fireflies one by one, or a few at a time, to observe how they coordinated the timing of their flashes.

Peleg quizzed Martin on the steps he'd taken to set up the cameras, and his responses underscored the differences between building theoretical models and testing them in the real world. "Checked to see they're at max ISO," Martin said. "They're on 1440 resolution, sixty frames per second, 360 video mode, pointing the same direction, calibrated." It would've been easy to leave a lens cap on, or botch the ISO setting, or fail to zip up the tent and let an entire night's subjects twinkle away into the foliage. Though their equipment was high tech, the experiments echoed some of the classic work done by the Bucks, who once netted scores of synchronous fireflies in Thailand and set them loose in their hotel room to see how they behaved.

Sarfati and Peleg are fairly confident that fireflies synchronize by following visual cues, but why males would flash as a collective when nature is full of males trying to distinguish themselves—consider the elaborate courtship dances of various birds, or the headbutting of rams—is another question. One hypothesis is that fireflies synchronize to reduce visual clutter, making it easier for females to locate them. But Peleg and Martin were interested in how fireflies adjust their flash frequencies, not why. And first, they needed to catch them.

Peleg walked along a raised boardwalk that ran perpendicular to the path where their tents were set up. She carried a butterfly net and made confident sweeps above the vegetation, netting fifteen fireflies in the time it took Martin to snag five. In the nets, the glow of the fireflies appeared green rather than white, like sparks from a strange fire. Although Peleg's data on *frontalis* was preliminary, her team had collected intriguing findings while observing *Photinus carolinus,* the synchronous species from the Great Smoky Mountains. "When we take them out of the swarm, they behave very differently," Peleg said. "In isolation, they behave one way. And when they're among their peers, they behave very differently." While *frontalis* synchronize in a steady rhythm for hours, *carolinus* flash in bursts, followed by communal pauses. But a lone *carolinus* doesn't follow the same pattern; its flashes and rests are sporadic. Once in the collective, it'll adjust its pauses to match the group's rhythm. In other words, the males aren't preprogrammed to a constant frequency; it's peer pressure that changes the timing mechanism in their brains.

The field of collective animal behavior is, in essence, a science of social influence. Explaining how one electron influences another is relatively easy, considering particles do not sense or think, or acknowledge rank, or select lifelong mates. Animals, on the other hand, evolved to make decisions within complex social contexts that are often riddled with loyalties and hierarchies. They see, hear, and smell; some detect pheromones, vibrations, or pressure. One animal might influence the decisions of another, but not vice versa, or the actions of animal A may affect those of animal B, which in turn circle back to influence animal A. (Now extend those simple influences between two pairs to populations of hundreds or thousands.) Sometimes animals cooperate when it's advantageous, but they can also be genetically selfish. They can also cheat, which distances them from the sorts of comprehensively predictable patterns that physicists are trained to search for.

Nevertheless, research increasingly suggests that distributed intelligence—the wisdom of the crowd—is a real thing. A 2013 study in *Science,* for example, found that golden shiners, a type of schooling fish, could detect gradients as a collective even as individual fish had no sense of the slope. A 2019 paper found that the same fish seemed to sense threats as a group, with the school growing jittery in the presence of a predator. The authors concluded that the school reorganized itself through the structure of its network; the individual fish weren't more sensitive to danger, but the system was.

"Humans tend to overly focus on the individual," said Iain Couzin, the director of the Max Planck Institute of Animal Behavior in Konstanz, Germany, and arguably the world's leading expert on collective behavior. "Even when we think about, you know, intelligence, we focus in on Einstein, we focus in on geniuses. We have this obsession with Steve Jobs as a genius, Einstein as a genius. . . . But in the vast majority of cases, the decisions we make are collective decisions, and the quality of those decisions is not correlated with individual IQ. It's more about the network of communication."

And while no one today attributes synchronicity to telepathy, collective behaviors do suggest that there's more to animal cognition than most of us acknowledge. Researchers have speculated that the incredible speeds at which birds pass information to one another within a flock could mean that individuals aren't merely responding to the positions and velocities of their closest neighbors, but instead anticipating what those neighbors may do in the future. It's a feat of mental time travel that's often considered the exclusive realm of humans.

The strength of Peleg's and Sarfati's work, Couzin said, is what it can teach us about the highly specific lives of individual species, as well as synchronicity as a model system. Its results may lead to the recognition of similar patterns in disparate biological systems—and to a fuller understanding of natural selection. "If one zooms out a little bit from fireflies and just thinks more broadly—think about our brains," he said. "Our brains are basically cells that are firing and influencing others around them that are firing. They're not flashing in terms of light, but they're flashing in terms of electricity and chemical pulses, and this creates these patterns and properties at higher levels."

Couzin, who writes about collective biological systems and the brain, has noted that certain ant colonies exhibit distinctive cycles of rest and activity that resemble neural wave patterns. He's also curious about whether criticality—which posits that systems can be perched at a point between order and disorder that yields remarkable properties—can explain rapid behavioral changes in flocks and schools. The concept has been integral to statistical physics for more than a century, but has recently been taken up by neuroscientists, who wonder whether it can account for various information-processing characteristics of the human brain. If criticality is behind the high-speed propagation of information across a flock, then it would be evidence of lossless information systems rooted in nature rather than in Silicon Valley.

Steven Strogatz, a professor of applied mathematics at Cornell and the author of *Sync: The Emerging Science of Spontaneous Order,* believes that the world's outstanding scientific mysteries, regardless of discipline, are all questions of collectives. Consider how tens of thousands of genes in a cell can interact to cause cancer. Or the fact that the human body has only about thirty thousand genes, yet can produce antibodies in response to an apparently limitless number of foreign substances. Even bacteria organize themselves to produce outcomes that extend beyond that of individual units. Antibiotic resistance, a terrifying public health challenge, is facilitated by colonies of bacteria swapping information across a genomic web.

"What happens when you have a group of about one hundred billion brain cells?" Strogatz asked me. "They start to have consciousness, they start to have emotions. An individual cell is not sad. But put enough of them together, and you get astonishing collective things that we call feelings and memories and perception."

By late June, Peleg and Sarfati had traded the heat and mosquitoes of Congaree for a brick building in Boulder, where tall windows in the stairwells frame spectacular views of the Flatirons. Along with Peleg's other postdocs and graduate students, Sarfati maintains a lab that includes an outdoor apiary alongside Boulder

Creek. There's a shared office jammed with desks and equation-covered whiteboards, a windowless cube with a fish tank housing a plant communication experiment, and a basement room stocked with equipment for handling honeybees: a floor sticky from sugary feed, four enormous jars of honey on a tall shelf, and a mesh barrier over the air vents to prevent the bees from escaping.

In another, more cramped room, which required visitors to don hooded beekeeper's jackets and gloves before entering, they keep X-ray equipment and, at times, bees. The bees emitted a quiet hum, belying the fact that there were approximately ten thousand of them. These particular honeybees hang in dense conical swarms and nimbly respond to stressors by flattening their swarm's shapes to resemble Frisbees. Peleg's team was x-raying them to understand how they did so. One little cone about three thousand bees strong hung from a wooden panel a few feet from our waists.

"The best way to connect it to the fireflies would be by trying to understand animal communication based on volatile signals in very large groups," Peleg said. "So with the fireflies, they create these flash patterns, and you can see how that signal is propagating throughout the entire swarm. And here we're dealing with either chemical communication with the bees, or mechanical communication. So they're holding hands, in a sense. Depending on how much force they exert on this cone, they're going to feel it." Then all three thousand get the message to reshuffle in a way that stabilizes the structure.

As for Sarfati, he'd gathered enough data on two hard drives to keep him busy with statistics for at least a year. The challenge now, he said as we sat on a bench in the dry mountain air, was less about getting good data than it was about asking interesting questions of it. If Peleg's experiments represented a bottom-up approach to understanding collective animal behavior by investigating what fireflies did when taken out of the swarm, then Sarfati's approach was top-down: he'd look at the swarm to find out how each individual works within the group.

His videos of fireflies would allow him to analyze three-dimensional reconstructions of firefly trajectories. A firefly's flash is obviously a signal, but could there be information encoded in their movements as well? The firefly known commonly as the Big Dipper, for instance, seems to communicate by movement as well as frequency, carving a distinctive J shape in the air as it flashes.

Sarfati planned to draw from the math of statistical physics and single-particle tracking to describe what the fireflies did on average, then zero in on the noise—the unknown bits—where he might find the individual animal reacting or thinking: evading a predator, for instance, or responding to a female.

In physics, he explained, some objects, like planets, have trajectories that are completely deterministic. On the other end of the spectrum, there's Brownian motion, which describes the fully random movements of microscopic particles. Sarfati's "thinking objects" fall somewhere in between, generating noise that's both confounding and potentially revelatory. Sarfati was fascinated by the idea that the physical laws governing inanimate objects might—with a few adjustments—also apply to animals; he thought the potential commonalities suggested that those principles contained deeper truths.

Of course, he could just as well discover the opposite. Studies into animals might augment—or at the very least complicate—the laws of physics. Albert Einstein himself pondered this possibility in 1949, in a letter in which he discussed a contemporary's studies of bees. "It is thinkable," he wrote, "that the investigation of the behaviour of migratory birds and carrier pigeons may some day lead to the understanding of some physical process which is not yet known."

Perhaps unsurprisingly, the kind of research that Peleg and Sarfati do is often celebrated less for what it tells us about the physical universe or animals, and more for what it might offer humans. Learning how bees and fireflies solve problems could have applications for communications systems, swarm robotics, and engineering, where it might aid in the development of far-out building materials that flex and adapt under strain. Sarfati's goals remain less selfish, if not less ambitious. "I'd just like to understand what they do," he said. "Just the fireflies themselves. I'd like to understand the firefly, for now."

FERRIS JABR

Brain Wave

FROM *The New York Times Magazine*

ON THE EVENING of October 10, 2006, Dennis DeGray's mind was nearly severed from his body. After a day of fishing, he returned to his home in Pacific Grove, Calif., and realized he had not yet taken out the trash or recycling. It was raining fairly hard, so he decided to sprint from his doorstep to the garbage cans outside with a bag in each hand. As he was running, he slipped on a patch of black mold beneath some oak trees, landed hard on his chin, and snapped his neck between his second and third vertebrae.

While recovering, DeGray, who was fifty-three at the time, learned from his doctors that he was permanently paralyzed from the collarbones down. With the exception of vestigial twitches, he cannot move his torso or limbs. "I'm about as hurt as you can get and not be on a ventilator," he told me. For several years after his accident, he "simply laid there, watching the History Channel" as he struggled to accept the reality of his injury.

Some time later, while at a fund-raising event for stem-cell research, he met Jaimie Henderson, a professor of neurosurgery at Stanford University. The pair got to talking about robots, a subject that had long interested DeGray, who grew up around his family's machine shop. As DeGray remembers it, Henderson captivated him with a single question: Do you want to fly a drone?

Henderson explained that he and his colleagues had been developing a brain-computer interface: an experimental connection between someone's brain and an external device, like a computer, robotic limb or drone, which the person could control simply by thinking. DeGray was eager to participate, eventually moving to

Menlo Park to be closer to Stanford as he waited for an opening in the study and the necessary permissions. In the summer of 2016, Henderson opened DeGray's skull and exposed his cortex—the thin, wrinkled, outermost layer of the brain—into which he implanted two 4-millimeter-by-4-millimeter electrode arrays resembling miniature beds of nails. Each array had one hundred tiny metal spikes that, collectively, recorded electric impulses surging along a couple of hundred neurons or so in the motor cortex, a brain region involved in voluntary movement.

After a recovery period, several of Henderson's collaborators assembled at DeGray's home and situated him in front of a computer screen displaying a ring of eight white dots the size of quarters, which took turns glowing orange. DeGray's task was to move a cursor toward the glowing dot using his thoughts alone. The scientists attached cables onto metal pedestals protruding from DeGray's head, which transmitted the electrical signals recorded in his brain to a decoder: a nearby network of computers running machine-learning algorithms.

The algorithms were constructed by David Brandman, at the time a doctoral student in neuroscience collaborating with the Stanford team through a consortium known as BrainGate. He designed them to rapidly associate different patterns of neural activity with different intended hand movements, and to update themselves every two to three seconds, in theory becoming more accurate each time. If the neurons in DeGray's skull were like notes on a piano, then his distinct intentions were analogous to unique musical compositions. An attempt to lift his hand would coincide with one neural melody, for example, while trying to move his hand to the right would correspond to another. As the decoder learned to identify the movements DeGray intended, it sent commands to move the cursor in the corresponding direction.

Brandman asked DeGray to imagine a movement that would give him intuitive control of the cursor. Staring at the computer screen, searching his mind for a way to begin, DeGray remembered a scene from the movie *Ghost* in which the deceased Sam Wheat (played by Patrick Swayze) invisibly slides a penny along a door to prove to his girlfriend that he still exists in a spectral form. DeGray pictured himself pushing the cursor with his finger as if it were the penny, willing it toward the target. Although he was physically incapable of moving his hand, he tried to do so with all his might. Brandman was ecstatic to see the decoder work as quickly as he had hoped. In thirty-seven seconds, DeGray gained control of the cursor and reached the first glowing dot. Within several minutes he hit dozens of targets in a row.

Only a few dozen people on the planet have had neural interfaces embedded in their cortical tissue as part of long-term clinical research. DeGray is now one of the most experienced and dedicated among them. Since that initial trial, he has spent more than eighteen hundred hours spanning nearly four hundred training sessions controlling various forms of technology with his mind. He has played a video game, manipulated a robotic limb, sent text messages and emails, purchased products on Amazon and even flown a drone—just a simulator, for now—all without lifting a finger. Together, DeGray and similar volunteers are exploring the frontier of a technology with the potential to fundamentally alter how humans and machines interact.

Scientists and engineers have been creating and studying brain-computer interfaces since the 1950s. Given how much of the brain's behavior remains a mystery—not least how consciousness emerges from three pounds of electric jelly—the aggregate achievements of such systems are remarkable. Paralyzed individuals with neural interfaces have learned to play simple tunes on a digital keyboard, control exoskeletons and maneuver robotic limbs with enough dexterity to drink from a bottle. In March, a team of international scientists published a study documenting for the first time that someone with complete, bodywide paralysis used a brain-computer interface to convey their wants and needs by forming sentences one letter at a time.

Neural interfaces can also create bidirectional pathways of communication between brain and machine. In 2016, Nathan Copeland, who was paralyzed from the chest down in a car accident, not only fist-bumped President Barack Obama with a robotic hand, he also experienced the tactile sensation of the bump in his own hand as the prosthesis sent signals back to electrodes in his brain, stimulating his sensory cortex. By combining brain-imaging technology and neural networks, scientists have also deciphered and partly reconstructed images from people's minds, producing misty imitations that resemble weathered Polaroids or smeared oil paintings.

Most researchers developing brain-computer interfaces say they are primarily interested in therapeutic applications, namely restoring movement and communication to people who are paralyzed or otherwise disabled. Yet the obvious potential of such technology and the increasing number of high-profile start-ups developing it suggest the possibility of much wider adoption: a future in which neural interfaces actually *enhance* people's innate abilities and grant them new ones, in addition to restoring those that have been lost.

In the history of life on Earth, we have never encountered a mind without a body. Highly complex cognition has always been situated in an intricate physical framework, whether eight suction-cupped arms, four furry limbs or a bundle of feather and beak. Human technology often amplifies the body's inherent abilities or extends the mind into the surrounding environment through the body. Art and writing, agriculture and engineering: All human innovations have depended on, and thus been constrained by, the body's capacity to physically manipulate whatever tools the mind devises. If brain-computer interfaces fulfill their promise, perhaps the most profound consequences will be this: Our species could transcend those constraints, bypassing the body through a new melding of mind and machine.

On a spring morning in 1893, during a military training exercise in Würzburg, Germany, a nineteen-year-old named Hans Berger was thrown from his horse and nearly crushed to death by the wheel of an artillery gun. The same morning, his sister, sixty miles away in Coburg, was flooded with foreboding and persuaded her father to send a telegram inquiring about her brother's well-being. That seemingly telepathic premonition obsessed Berger, compelling him to study the mysteries of the mind. His efforts culminated in the 1920s with the invention of electroencephalography (EEG): a method of recording electrical activity in the brain using electrodes attached to the scalp. The oscillating patterns his apparatus produced, reminiscent of a seismograph's scribbling, were the first transcriptions of the human brain's cellular chatter.

In the following decades, scientists learned new ways to record, manipulate, and channel the brain's electrical signals, constructing ever-more-elaborate bridges between mind and machine. In 1964, José Manuel Rodríguez Delgado, a Spanish neurophysiologist, brought a charging bull to a halt using radio-controlled electrodes embedded in the animal's brain. In the 1970s, the University of California Los Angeles professor Jacques Vidal coined the term brain-computer interface and demonstrated that people could mentally guide a cursor through a simple virtual maze. By the early 2000s, the Duke University neuroscientist Miguel Nicolelis and his collaborators had published studies demonstrating that monkeys implanted with neural interfaces could control robotic prostheses with their minds. In 2004, Matt Nagle, who was paralyzed from the shoulders down, became the first human to do the same. He further learned how to use his thoughts alone to play Pong, change channels on a television, open emails, and draw a circle on a computer screen.

Since then, the pace of achievements in the field of brain-computer interfaces has increased greatly, thanks in part to the rapid development of artificial intelligence. Machine-learning software has substantially improved the efficiency and accuracy of neural interfaces by automating some of the necessary computation and anticipating the intentions of human users, not unlike how your phone or email now has A.I.-assisted predictive text. Last year, the University of California San Francisco neurosurgeon Edward Chang and a dozen collaborators published a landmark study describing how a neural interface gave a paralyzed thirty-six-year-old man a voice for the first time in more than fifteen years. Following a car crash and severe stroke at age twenty, the man, known as Pancho, lost the ability to produce intelligible speech. Over a period of about twenty months, 128 disk-shaped electrodes placed on top of Pancho's sensorimotor cortex recorded electrical activity in brain regions involved in speech processing and vocal tract control as he attempted to speak words aloud. A decoder associated different patterns of neural activity with different words and, with the help of language-prediction algorithms, eventually learned to decipher 15 words per minute with 75 percent accuracy on average. Although this is only a fraction of the rate of typical speech in English (140 to 200 words a minute), it is considerably faster than many point-and-click methods of communication available to people with severe paralysis.

In another groundbreaking study published last year, Jaimie Henderson and several colleagues, including Francis Willett, a biomedical engineer, and Krishna Shenoy, an electrical engineer, reported an equally impressive yet entirely different approach to communication by neural interface. The scientists recorded neurons firing in Dennis DeGray's brain as he visualized himself writing words with a pen on a notepad, trying to re-create the distinct hand movements required for each letter. He mentally wrote thousands of words in order for the system to reliably recognize the unique patterns of neural activity specific to each letter and output words on a screen. "You really learn to hate M's after a while," he told me with characteristic good humor. Ultimately, the method was extremely successful. DeGray was able to type up to ninety characters or eighteen words a minute—more than twice the speed of his previous efforts with a cursor and virtual keyboard. He is the world's fastest mental typist. "Sometimes I get going so fast it's just one big blur," he said. "My concentration gets to a point where it's not unusual for them to remind me to breathe."

Achievements in brain-computer interfaces to date have relied on a mix of invasive and noninvasive technologies. Many scientists in the field, including those who work with DeGray, rely on a surgically embedded array of spiky electrodes produced by a Utah-based company, Blackrock Neurotech. The Utah Array, as it's known, can differentiate the signals of individual neurons, providing more refined control of connected devices, but the surgery it requires can result in infection, inflammation and scarring, which may contribute to eventual degradation of signal strength. Interfaces that reside outside the skull, like headsets that depend on EEG, are currently limited to eavesdropping on the collective firing of groups of neurons, sacrificing power and precision for safety. Further complicating the situation, most neural interfaces studied in labs require cumbersome hardware, cables and an entourage of computers, whereas most commercially available interfaces are essentially remote controls for rudimentary video games, toys, and apps. These commercial headsets don't solve any real-world problems, and the more powerful systems in clinical studies are too impractical for everyday use.

With this problem in mind, Elon Musk's company Neuralink has developed an array of flexible polymer threads studded with more than three thousand tiny electrodes connected to a bottlecap-size wireless radio and signal processor, as well as a robot that can surgically implant the threads in the brain, avoiding blood vessels to reduce inflammation. Neuralink has tested its system in animals and has said it would begin human trials this year.

Synchron, which is based in New York, has developed a device called a Stentrode that doesn't require open-brain surgery. It is a four-centimeter, self-expanding tubular lattice of electrodes, which is inserted into one of the brain's major blood vessels via the jugular vein. Once in place, a Stentrode detects local electric fields produced by nearby groups of neurons in the motor cortex and relays recorded signals to a wireless transmitter embedded in the chest, which passes them on to an external decoder. In 2021, Synchron became the first company to receive FDA approval to conduct human clinical trials of a permanently implantable brain-computer interface. So far, four people with varied levels of paralysis have received Stentrodes and used them, some in combination with eye-tracking and other assistive technologies, to control personal computers while unsupervised at home.

Philip O'Keefe, sixty-two, of Greendale, Australia, received a Stentrode in April 2020. Because of amyotrophic lateral sclerosis (ALS), O'Keefe can walk only short distances, cannot move his left arm, and is losing the ability to speak clearly. At first, he explained, he had to concentrate intensely on the imagined movements required to operate the system—in his case, thinking about moving his left ankle for different lengths of time. "But the more you use it, the more it's like riding a bike," he said. "You get to a stage where you don't think so hard about the movement you need to make. You think about the function you need to execute, whether it's opening an email, scrolling a web page, or typing some letters." In December, O'Keefe became the first person in the world to post to Twitter using a neural interface: "No need for keystrokes or voices," he wrote by mind. "I created this tweet just by thinking it. #helloworldbci"

Thomas Oxley, a neurologist and the founding CEO of Synchron, thinks future brain-computer interfaces will fall somewhere between LASIK and cardiac pacemakers in terms of their cost and safety, helping people with disabilities recover the capacity to engage with their physical surroundings and a rapidly evolving digital environment. "Beyond that," he says, "if this technology allows anyone to engage with the digital world better than with an ordinary human body, that is where it gets really interesting. To express emotion, to express ideas—everything you do to communicate what is happening in your brain has to happen through the control of muscles. Brain-computer interfaces are ultimately going to enable a passage of information that goes beyond the limitations of the human body. And from that perspective, I think the capacity of the human brain is actually going to increase."

There is no technology yet that can communicate human thoughts as fast as they occur. Fingers and thumbs will never move quickly enough. And there are many forms of information processing better suited to a computer than to a human brain. Oxley speculated about the possibility of using neural interfaces to enhance human memory, bolster innate navigational skills with a direct link to GPS, sharply increase the human brain's computational abilities, and create a new form of communication in which emotions are wordlessly "thrown" from one mind to another. "It's just the beginning of the dawn of this space," Oxley said. "It's really going to change the way we interact with one another as a species."

Frederic Gilbert, a philosopher at the University of Tasmania, has studied the ethical quandaries posed by neurotechnology for more than a decade. Through in-depth interviews, he and other ethicists have documented how some people have adverse reactions to neural implants, including self-estrangement, increased impulsivity, mania, self-harm and attempted suicide. In 2015, he traveled to Penola, South Australia, to meet Rita Leggett, a fifty-four-year-old patient with a very different, though equally troubling, experience.

Several years earlier, Leggett participated in the first human clinical trial of a particular brain-computer interface that warned people with epilepsy of imminent seizures via a handheld beeper, giving them enough time to take a stabilizing medication or get to a safe place. With the implant, she felt much more confident and capable and far less anxious. Over time, it became inextricable from her identity. "It was me, it became me," she told Gilbert. "With this device I found myself." Around 2013, NeuroVista, the company that manufactured the neural interface, folded because it could not secure new funding. Despite her resistance, Leggett underwent an explantation. She was devastated. "Her symbiosis was so profound," Gilbert told me, that when the device was removed, "she suffered a trauma."

In a striking parallel, a recent investigation by the engineering magazine *IEEE Spectrum* revealed that, because of insufficient revenues, the Los Angeles–based neuroprosthetics company Second Sight had stopped producing and largely stopped servicing the bionic eyes they sold to more than 350 visually impaired people around the world. At least one individual's implant has already failed with no way to repair it—a situation that could befall many others. Some patients enrolled in clinical trials for Second Sight's latest neural interface, which directly stimulates the visual cortex, have either removed the device or are contemplating doing so.

If sophisticated brain-computer interfaces eventually transcend medical applications and become consumer goods available to the general public, the ethical considerations surrounding them multiply exponentially. In a 2017 commentary on neurotechnology, the Columbia University neurobiologist Rafael Yuste and twenty-four colleagues identified four main areas of concern: augmentation; bias; privacy and consent; and agency and identity. Neural implants sometimes cause disconcerting shifts in patients' self-perception. Some have reported feeling like "an electronic doll" or developing a blurred sense of self. Were someone to commit a crime and blame an implant, how would the legal system determine fault? As neural interfaces and artificial intelligence evolve, these tensions will probably intensify.

All the scientists and engineers I spoke to acknowledged the ethical issues posed by neural interfaces, yet most were more preoccupied with consent and safety than what they regarded as far-off or unproven concerns about privacy and agency. In the world of academic scientific research, the appropriate future boundaries for the technology remain contentious.

In the private sector, ethics are often a footnote to enthusiasm, when they are mentioned at all. As pressure builds to secure funding and commercialize, spectacular and sometimes terrifying claims proliferate. Christian Angermayer, a German entrepreneur and investor, has said he is confident that everyone will be using brain-computer interfaces within twenty years. "It is fundamentally an input-output device for the brain, and it can benefit a large portion of society," he posted on LinkedIn last year. "People will communicate with each other, get work done and even create beautiful artwork, directly with their minds." Musk has described the ultimate goal of Neuralink as achieving "a sort of symbiosis with artificial intelligence" so that humanity is not obliterated, subjugated, or "left behind" by superintelligent machines. "If you can't beat 'em, join 'em," he once said on Twitter, calling it a "Neuralink mission statement." And Max Hodak, a former Neuralink president who was

forced out of the company, then went on to found a new one called
Science, dreams of using neural implants to make the human sen-
sorium "directly programmable" and thereby create a "world of
bits": a parallel virtual environment, a lucid waking dream, that
appears every time someone closes their eyes.

Today, DeGray, sixty-eight, still resides in the Menlo Park assisted-
living facility he chose a decade ago for its proximity to Stanford. He
still has the same two electrode arrays that Henderson embedded
in his brain six years ago, as well as the protruding metal pedestals
that provide connection points to external machines. Most of the
time, he doesn't feel their presence, though an accidental knock
can reverberate through his skull as if it were a struck gong. In his
everyday life, he relies on round-the-clock attention from caregiv-
ers and a suite of assistive technologies, including voice commands
and head-motion tracking. He can get around in a breath-operated
wheelchair, but long trips are taxing. He spends much of his time
reading news articles, scientific studies, and fiction on his computer. "I really miss books," he told me. "They smell nice and feel
good in your hands."

DeGray's personal involvement in research on brain-computer
interfaces has become the focus of his life. Scientists from Stanford visit his home twice a week, on average, to continue their
studies. "I refer to myself as a test pilot," he said. "My responsibility
is to take a nice new airplane out every morning and fly the wings
off of it. Then the engineers drag it back into the hangar and fix it
up, and we do the whole thing again the next day."

Exactly what DeGray experiences when he activates his neural
interface depends on his task. Controlling a cursor with attempted
hand movements, for example, "boils the whole world down to an
Etch A Sketch. All you have is left, right, up and down." Over time,
this kind of control becomes so immediate and intuitive that it
feels like a seamless extension of his will. In contrast, maneuvering
a robot arm in three dimensions is a much more reciprocal process: "I'm not making it do stuff," he told me. "It's working with
me in the most versatile of ways. The two of us together are like a
dance."

No one knows exactly how long existing electrode arrays can
remain in a human brain without breaking down or endangering
someone's health. Although DeGray can request explantation at

any time, he wants to continue as a research participant indefinitely. "I feel very validated in what I'm doing here," he said. "It would break my heart if I had to get out of this program for some reason."

Regarding the long-term future of the technology in his skull, however, he is somewhat conflicted. "I actually spend quite a bit of time worrying about this," he told me. "I'm sure it will be misused, as every technology is when it first comes out. Hopefully that will drive some understanding of where it should reside in our civilization. I think ultimately you have to trust in the basic goodness of man—otherwise, you would not pursue any new technologies ever. You have to just develop it and let it become monetized and see where it goes. It's like having a baby: You only get to raise them for a while, and then you have to turn them loose on the world."

The Provincetown Breakthrough

FROM *Wired*

IT IS JULY 10, 2021, a Saturday, and Sean Holihan is on a short flight down the East Coast of the United States. The nonprofit strategist and his partner are coming home from Provincetown, Massachusetts, where they spent the week with friends they hadn't seen since the pandemic began. Everyone was vaccinated, and the state had dropped its mask mandate, so the holiday felt gloriously normal. The group rented cottages and hung out together, sharing brunch and cocktails, hitting the beach in the afternoon and dinner and shows at night. Provincetown was its old delirious self: sixty thousand visitors turned a mile-long portion of Commercial Street, the main thoroughfare, into an impromptu parade, and crammed into nightclubs so crowded that you had to slide skin to skin to get outside for some air. Holihan is exhausted, but happy.

On the same day, in New York City, a data scientist named Michael Donnelly is making plans with friends who are driving back from Provincetown. He and his husband try to go every summer, but this year things booked up crazy fast. They're all planning to meet up tonight when everyone arrives. Donnelly maintains a COVID analysis site as a hobby and has been the nerd node for his friends when they need information. He's looking forward to taking the night off.

On Cape Cod, Theresa Covell has just gotten back from her first vacation since the pandemic began. Covell is an assistant public health nurse for Barnstable County, the jurisdiction that stretches all along the arm of the Cape, from the shoulder joint at the Bourne Canal to the wrist curve that shelters Provincetown from the ocean.

She and her colleagues have spent 2021 grinding—tracking cases, running vaccine clinics, trying to manage the emergency in a place that is low on revenue when the tourists are gone and short of housing and services once they arrive. When she left for her time off, the COVID curve was bending down. In all of June, Provincetown didn't see a single positive case. But recently a local health organization has reported a surge among vaccinated people. *That's new*, Covell thinks.

Holihan logs on to the airplane Wi-Fi, and he feels his phone vibrate. It's a friend who checked out early this morning from their rental, complaining of a summer cold. He has gotten home and taken a COVID test. The text shows the bright double lines of a positive result. Holihan's first reaction is disbelief: The authorities had said they were safe. His second is dread. *I'm on a plane*, he thinks. *Am I going to give this to other people?*

Donnelly's phone lights up. His carload of friends just got word that someone they know tested positive, and they've pulled off the highway in Connecticut in search of rapid tests. They are all vaccinated, but three of the five soon test positive. Their plans to meet are off. By the end of the day, another twelve people who were in Provincetown tell Donnelly they've contracted COVID. *This isn't supposed to happen*, he thinks. It feels like the floor is falling out from under him.

If you even remember the Provincetown outbreak a year ago—which, in Pandemic Time, probably feels like a century—this may be what you know about it: The shots had been available for seven months. The Centers for Disease Control and Prevention had said that vaccinated Americans could take their masks off. There was a delicious anticipatory buzz of life returning to normal, and then the ice-bucket shock of discovering hot vax summer was over before it started. The Delta variant caused breakthrough infections and illness even in vaccinated people, and the Provincetown outbreak was the proof.

Maybe you remember the disappointment of mask recommendations coming back, or the whiff of homophobia that floated through some of the coverage, or the sense that a summer capital for artists and queer people had been made responsible for the Delta variant rather than being its accidental host. ("How do we stop the press portraying us as a leper colony?" a local business owner asked on the town manager's Facebook page.)

Whatever you remember, the actual story is this. The partyers
in Provincetown didn't spread the virus; they, and their allies, controlled it. On the fly, they created a model for how a community
can organize against a disease threat. Even a year later, it is worth
looking back at what they did—not just because COVID has not left
us but also because other pandemics will come. Much of the US
response to COVID has been fractured, hostile, or self-sabotaging.
Provincetown was "a huge success story," says William Hanage, co-
director of Harvard's Center for Communicable Disease Dynamics,
who helped analyze the outbreak. "It should have been a message:
We can avoid large outbreaks, if we want to."

Holihan and his partner reached their home in Washington, DC,
and found some rapid tests tucked under their doormat—a gift
from the friend who had texted his positive result. They both took
them right away. Holihan's test popped positive immediately. They
masked up and maneuvered awkwardly around their apartment,
trying to figure out where they could separately eat and sleep. The
next morning, feverish and sweating, Holihan walked to a pop-up
clinic for a PCR test, then went home to isolate.

The next day, Monday, he emailed his office—he is the state
legislative director for a gun-violence-prevention organization—to
say he wouldn't be in. He was already feeling better, but when his
test result arrived, it was positive. Of course.

"At that point I started texting everyone I'd come in contact
with over the week," he says. Realizing how many people visit Provincetown from across the country, he posted about being infected
on Twitter and Instagram too. DMs flowed back, from people who
thought they'd picked up some summer crud as they traveled.
"They thought they were fine," he says. "Then they tested themselves, and it turned out they also had COVID."

One of the people Holihan texted was Donnelly. This might
seem odd, because Donnelly isn't an epidemiologist. He is a policy
geek who has done macroeconomic forecasting at the Federal Reserve Board and data analysis at Spotify and Facebook. But since
early 2020, Donnelly had also been applying his skills to forecasting what COVID might do in the US, a way of making sense for
himself of the data flowing from other countries and explaining to
others why they ought to be more worried than they were. "Essentially, I wanted to convince my friends it was bad," he says.

Donnelly's analyses, which he initially published on Medium, had been solid. He had foreseen that federal action would be needed two days before President Donald Trump declared a national emergency. He had warned that New York City would have to shut down six days before Governor Andrew Cuomo announced that the whole state would be put "on pause." That prediction led to a consulting gig with New York state (forecasting possible case counts, bed needs, and ventilator orders) and then to founding a site called CovidOutlook.info, a home for reports and predictions that he spun up with Michael LeVasseur, an epidemiologist at Drexel University.

So by the time the Delta variant began creeping through Provincetown, Donnelly was an informal but thoroughly informed expert in what COVID was doing in the US. "I had been tracking variants over the previous six months and, broadly, thought concerns about them were overblown," he says. When his friends started testing positive, he was surprised and nettled. He didn't like being wrong.

Rumors about people testing positive were zipping through group chats: most of this house, everyone in that cottage; the Pennsylvania group, the California group, that couple from DC; ten people positive, or fifteen, or twenty-five. Text by text, Donnelly began verifying the stories, asking people about the symptoms they had and the tests they had taken, when they were vaccinated and which shot they got, and all the details of their visits to Provincetown—where they stayed, who they hung out with, which bars and restaurants and shows they went to. He started collecting information on Saturday afternoon, and by Monday he had more than fifty names in a spreadsheet.

The list represented a shocking number of breakthrough infections for a young, healthy, affluent population, a group that should have been at the lowest risk. Donnelly felt an itch to do a study, but LeVasseur persuaded him to turn the project over to a bigger institution than their team of two. Donnelly got in touch with Demetre Daskalakis, the former head of the infectious disease programs in New York City's health department, who was now at the CDC. On Monday night, Donnelly texted, offering the spreadsheet. Daskalakis asked for it immediately.

Within twenty-four hours, Daskalakis set up calls between Donnelly, the CDC, and the Massachusetts health department. By the end of the week, the agencies had created a task force, set up a phone number and an email for people to self-report, reached out to other states that visitors had gone home to, and gotten mobile testing units rolling toward Provincetown. "It's the most accelerated response I've ever seen in public health," Daskalakis says. "And Michael pretty much started that outbreak investigation himself."

This ought to be obvious, but to make it clear: Everyone Donnelly interviewed, those who visited Provincetown and came away with a COVID infection, is gay. That was why they were in that place in one of its biggest tourist weeks of the year. But people don't go to Provincetown just to party—they go for community. "Even if you live in LGBT-friendly neighborhoods, you're still a minority," says Rob Anderson, a former journalist who moved to the town a decade ago and owns a Commercial Street restaurant called The Canteen with his partner. "Even in New York, you can be walking down the street and get called a fag. When you come to Provincetown, you just get to be yourself. You feel normal, for the first time in your life."

The physical setting helps with that. Provincetown is remote—it lies at the dead end of miles of two-lane highway—and pretty in an unthreatening way, made up of low, shingled buildings surrounded by tidal marshes and soft ocean light. But its social norms help too. It is a place that takes openness to sex and gender expression as a basic social contract. That openness might show up as a guy wearing heels and fairy wings to the corner store, or a mom bringing her kids cross-country to the beach because every child there will have queer parents—or the town collectively accepting that tens of thousands of smooching, shouting people will flood the streets and clubs for the themed holidays that segment the summer: Memorial Day for young lesbians, July Fourth for gym guys, Bear Week right afterward for big, hairy men, special weeks for Black queer men and women, a raucous costumed Carnival to close out the summer.

Despite the partying, there's a shadow of trauma present in Provincetown, an acknowledgment of the long grief of the HIV pandemic—which was identified in the US forty years, minus a few weeks, before Delta came to town. Provincetown has been a queer community for so long that AIDS is not past history there, even though it has been survivable for more than twenty years and preventable for just about ten. Before good treatments were available, some men who were infected fled there to escape stigma; across from the towering Pilgrim Monument, there's a memorial to the lost, a massive slab of quartzite carved to resemble the surface of the ocean. In a way, Provincetown's sex-positive culture owes its existence to the health-focused practices imposed by HIV: not just staying alert to the risk of infection and practicing safe sex, but also getting tested regularly and disclosing your status when it changes.

"We've had to develop social norms and expectations to share our risks and exposures," says Donnelly, who is thirty-seven, born after the first, worst years of HIV. "I don't want to be Pollyannaish about it: This is still work. We're not perfect at it."

Thus many Provincetown visitors and residents were primed, the way a vaccination primes the body to fight a later infection, to recognize that Delta was spreading among them and to be very public about it. People who realized they'd been exposed in the July Fourth week went further than simply admitting to Donnelly that they tested positive. They began doing contact tracing on themselves and looked for professionals to give the information to.

One of them was Daskalakis. "I got emails, and the emails went, 'Hi, my name is X. On Monday I was here, on Tuesday I was here, on Thursday I had dinner with this person,'" he recalls. "It was amazing. Other CDC folks will tell you: It was unlike any other group they've dealt with in terms of getting information."

The men identifying themselves were fully aware that there might be a cost to doing so. Some of the residents remember, and just about everyone has heard, about the ways in which HIV-positive men were blamed for their own illnesses. Speaking up about COVID meant risking that again—both from the wider world (right-wing media were vicious) and within the gay community. Holihan's tweet about his positive result didn't only draw encouraging DMs. "I had old friends reach out and say I had been irresponsible, almost like a little bit of slut-shaming," he recalls.

Theresa Covell returned from her days off to find her boss, Deirdre Arvidson, and her colleague Maurice Melchiono—the entirety of the Barnstable County public-health nursing team—turning back to the tasks they'd been doing for a year: Receiving reports of positive cases from the state. Identifying infected people and calling and counseling them. Making sure those people were isolating and finding out whether they had a workplace to be notified or kids who needed care. Getting them help if they needed pulse oximeters, grocery deliveries, a separate place to sleep. Calling again, to make sure they were managing. Calling again after that, to make sure they had recovered. Calling and calling and calling, from lists that had new names added every day.

Covell and her coworkers were stunned by the spread. They knew that some bars and clubs in town were checking visitors' vaccination cards, and that most locals had gotten the shots. But this new wave of COVID didn't seem to care. "First it was the gay and bisexual population, and then it was the seasonal workers, and then it became residents and then schoolkids, and it just continued to grow," Melchiono says.

The spiraling case count made clear how much individual behavior and local conditions mattered to the transmission of the virus. A tropical storm had churned up the coast that holiday week, and the weather had been cold and rainy enough to drive people indoors instead of enticing them onto beaches and balconies. And though federal guidance said vaccinated people were safe indoors and face-to-face, that didn't account for the unique context of Provincetown—especially its thousands of seasonal workers, some unvaccinated or undervaccinated, bunking in campgrounds and crowded temporary housing. The town government reacted by recommending masks on July 19 and mandating them on July 25, but the outbreak tore through the workforce.

Covell's team had only enough resources and jurisdiction to investigate within Barnstable County. For tracking the people who'd left the Cape, there was a bigger, better-funded effort—the Community Tracing Collaborative, a 4,000-person corps created by the state health department and the global nonprofit Partners in Health. Its size suggested the scope of the job. COVID isn't a disease that's amenable to traditional contact tracing. It doesn't transmit only one-to-one, the way Ebola, monkeypox, leprosy, and HIV do, but one-to-many as well. Boston itself had been host to one of the largest such events, a super-spreader biotech conference in February 2020 that over months caused more than 330,000 cases worldwide. There would have been no point in trying to understand which person at that conference infected which other attendee. But if officials had been able to warn all the attendees to isolate, they might have prevented later generations of infection.

That was what the Community Tracing Collaborative undertook for Provincetown. It was a practice of pattern recognition, using data analysis tools to map the relationships between people and the places they had gone and the other people who might have been present at the same time. It required a bending of the normal rules of disease investigations, which strictly define what constitutes a case and what qualifies as exposure—and it asked investigators to look at both where the risks had been and where they might go next.

This simultaneous tracing of people and gatherings was a newer approach in the US—the method was copied from COVID strategies in Japan—and it wasn't easy. If you diagram the transmission of a disease that goes person-to-person, it looks like a family tree. In Provincetown, it looked like an overgrown forest. "There were so many overlapping interactions, across three different streets, in upwards of twenty different locations—so it was very difficult to pinpoint where someone was actually exposed," says Perri Kasen, a management consultant who joined the Community Tracing Collaborative in 2020 and became one of the three lead investigators for the Provincetown outbreak. Among the hundreds of cases detected in the outbreak in the first half of July, contact tracers could identify only *six people* for whom it was reasonably certain that one had infected the other. But, as Kasen says, "you don't necessarily need confirmatory evidence to act."

On July 27, the CDC did act. In a bombshell media briefing, director Rochelle Walensky announced gloomily that vaccinated people should go back to wearing masks indoors, especially in schools and around the vulnerable. New data, she said, had shown that when vaccinated people developed breakthrough infections from the Delta variant, they carried the same amount of virus as infected people who had never been vaccinated, and could pass the virus to others. "It is not a welcomed piece of news," she said. "This new data weighs heavily on me." Three days later, the agency released the data in its weekly journal. It was an analysis of an outbreak in "a town in Barnstable County" during "multiple summer events and large public gatherings." That day, there were 105,120 new COVID cases reported in the United States—six times as many as on July 1.

*

As the warm months went on and America mourned the loss of hot vax summer, Bronwyn MacInnis tried to figure out exactly how the outbreak happened. She is in charge of pathogen genomic surveillance at the Broad Institute, a private research facility shared between MIT and Harvard. For more than a year, her team had been sequencing the viruses collected in COVID tests, helping the state understand its local epidemic. Like everyone else, she had started to relax as cases declined. Fewer samples came in; the vaccines seemed to be doing their job. And like everyone else, she had been shocked to hear about the Provincetown cluster. She got the call while she was riding her bike through Harvard Square. She remembers thinking: *Maybe I'd better pull over.*

She and her team zeroed in on the tiny mutations that occur when SARS-CoV-2 reproduces, both within a single person and also as it passes from one person to another. "Asking humans where they've been and who they have been in contact with can be really complicated," MacInnis says. "But the viruses can tell you that information in black and white." Whenever the researchers encountered a gap in the genomic narrative, they used data from state epidemiologists and contact tracers to fill it in.

Piece by piece, they built their model. By October they could define the outbreak's full size: 1,098 people infected in the Provincetown area in July. Based on subtle genetic differences, they determined that varieties of the Delta variant were introduced to the town more than forty times that month. Five of those introductions led to small clusters of cases, and one was responsible for most of the outbreak—83 percent. (This could have been one person or a family or other small group.)

Next, MacInnis's team compared the signatures of the Provincetown strains with genomes from across the country. Though people infected in the outbreak had come from at least twenty other states and Washington, DC, those genetic signatures were almost nowhere to be found. The outbreak did not amplify across the US. Instead, it fizzled. Eight people got sick enough to be hospitalized. No one died. By the middle of September, the Provincetown strains accounted for no more than 0.1 percent of cases nationwide.

"It was almost a moment of tears to see how limited the transmission appeared to be," MacInnis said. "If we hadn't had the sequences of those viruses, I have no doubt that the public narrative about this outbreak would have been very different." The lightning group chats, the visitors offering up their data, the frantic phone-calling by the Barnstable County nursing team, the data massaging that Kasen and the contact tracers did—it had all helped. Even though Delta swamped the country anyway.

No occurrence of a disease is fortunate. Still, it was an extraordinary piece of pandemic luck that the first US explosion of Delta took place in a community so willing to offer up its lives for examination by strangers. It was a second piece of luck that those stories were told in a state that had the infrastructure to receive them, alongside a research institute equipped and eager to trace what the virus did next. None of that, though was foreordained. Another reality could have played out just as easily: The Provincetown visitors arrive home, notice what they think are summer colds, and don't test themselves. They tell no one. The authorities are slow to notice the outbreak. The contact tracing and genomic analyses take longer to spin up, and people spend weeks or months under the false impression that they're fully protected. In this reality, more people would have gotten sick. Almost certainly, some would have died.

Those people were saved, whoever they were, because the men who visited Provincetown took on the burden of going public, and the nurses and contact tracers and scientists transformed their information into reasons to act. In a pandemic marked mostly by how much people have arrayed themselves against each other, they chose to act for others. In the never-ending battle between plagues and people, they chose—as anyone can—not to be on the side of the plague.

True Grit

FROM *The Atavist Magazine*

THE WILD HORSES all have names. Ronald, for example, and Becky and Clyde. The names sound mundane, even for horses, but each is something like a badge of honor. For years now, the people of Cedar Island, North Carolina, have named each foal born to the local herd of mustangs after the oldest living resident who hasn't already had a horse named for them. Every island family of long standing has this connection to the herd.

Cedar Island, located in a pocket of North Carolina known as Down East, is what passes for remote in the continental United States these days. Though it's only forty miles as the gull flies from the Cape Hatteras area, with its tourists and mortgage brokers, its restaurants with names like Dirty Dick's Crab House, Cedar Island remains a place with only a scattering of people and businesses, where you can't be certain of finding a restaurant meal—not so much as a plate of hush puppies—on a Sunday evening. Upon arrival you might not notice that Cedar Island is an island at all. Crossing the soaring Monroe Gaskill Memorial Bridge, which connects it to the mainland, what you pass over is easily mistaken for another of the region's sleepy, curlicue rivers. In fact, this is the Thorofare, a skinny saltwater channel connecting the Pamlico Sound to the north and the Core Sound to the south. The Pamlico is one of the largest embayments on the US coastline, while the Core is narrow and compact. Cedar Island stands between them, and all three are hemmed in by the Outer Banks.

I've just written that Cedar Island separates two sounds, and on maps this is true. Reality is less decisive. Swaths of the small island are sometimes underwater, depending on wind, tide, and season—in particular, hurricane season.

The shifting, amphibious nature of Cedar Island was never more apparent than on the morning of September 6, 2019. Under the whirling violence of Hurricane Dorian, maps lost all meaning. The Pamlico and Core Sounds joined to become a single angry body of water, shrinking Cedar Island to a fraction of its acreage. It was no longer separated from the mainland by the thin blue line of the Thorofare, but by nearly six miles of ocean.

Most of the 250 or so people living on the island were safe, their homes built on a strip of not-very-high high ground precisely to weather the wrath of hurricanes. The wild horses—forty-nine in all—were in much deeper trouble.

There were also some cows. The cows did not have names.

There is no such thing as a truly wild cow. While Cedar Island's cattle range more or less freely, the technical term for them is *feral*—they are the descendants of escapees from domestication. The island's mustangs are feral, too, but while visitors often come to Cedar Island solely in hopes of seeing the Banker horses, as the area's herds are known, next to no one makes a special trip to photograph the "sea cows."

The cows are striking to look at, though. While they vary in color, many have a bleached-blond coat, blending in with the pale sand and the glare of the sun on Cedar Island's hammerhead northern cape, where both cattle and horses roam. Tourists are happy to see the cows, just not *as* happy as they are to see the horses. Here and across America, a mustang—mane flowing, hooves pounding the earth—is an embodiment of beauty and freedom. Cows are not.

For Cedar Islanders, the cows are part of what makes their home distinctive, a fond and familiar part of the community and its history. In fact, the cattle have been on the island far longer than the mustangs, who were transferred from the more famous Shackleford Banks herd three decades ago. But the relationship people on the island have with horses is different from the one they have with cows, in much the same way it is for people nearly everywhere.

"This used to be horse country," said Priscilla Styron, who has lived on or near Cedar Island for thirty years and works at its ferry terminal. "Everybody rode, they had pony pennings, they had all kinds of stuff. Everybody was always riding horses." As for the cows, there was a time not so long ago when an islander might round one up from the beach, take it home to graze and fatten up, then butcher it for meat.

As Hurricane Dorian approached Cedar Island, no one troubled themselves about either kind of animal. One islander, who called himself a "simple country boy" and asked not to be named, scoffed at the idea that wild creatures would brook being corralled and taken off-island to wait out the storm. Not that anyone thought that was needed, according to Styron. "They usually protect themselves. You don't have to worry about them," she said. "They can sense more than we can." Cedar Island had never lost more than one or two members of its wild herds to a storm—and Down East sees more than its fair share of those.

In 2019, there were perhaps a couple dozen cattle on the island— no one knew for sure, because no one was keeping count, not even residents who were fond of their bovine neighbors. For at least some of the cows, Dorian was nothing new. Few cows in America live longer than six years; many are slaughtered much younger. A Cedar Island cow, on the other hand, stands a good chance of living into its teens, and might even see its thirtieth birthday. A cow that was twenty years old in 2019 would have had close encounters with at least ten hurricanes: Dennis, Floyd, Isabel, Alex, Ophelia, Arthur, Matthew, Florence, and two named Irene. The herd could look to its elders for guidance.

Biologists have only recently recognized that cows have complex social behaviors, involving depths of comprehension that we might not expect of animals stereotyped as grungy, placid, and dull-witted. A feral herd, for example, will organize nurseries by dividing calves into age groups, each usually overseen by one adult cow while the rest go out to graze. For this to work, the sitters need to understand that their role is to look after calves that are not their own, even if it means settling for low-grade fodder while others enjoy greener pastures. The calves have to grasp that they are under vigilance despite their mothers being out of sight.

No one documented how the cows responded as Dorian approached, but Mónica Padilla de la Torre, an evolutionary biologist, can give us a good idea. "They usually are not afraid of storms. They like storms," Padilla said. "They like to be cool. They like shade. They appreciate when the rain comes."

Even before the hurricane loomed on the southern horizon, the herd likely began to move—with that usual cattle slowness, that walking-on-the-moon gait—toward shelter. In the era before hurricanes were tracked by satellites and weather radar, cows were a useful predictor that one was coming. The migration, Padilla said, would have been initiated by the herd's leaders. Cattle violently clash to establish a pecking order, and once that's settled, a benign dictatorship ensues. Leaders are granted the best places to eat and the best shade to lie in, and they make important decisions—like when to retreat to high ground in the face of a storm.

For Cedar Island's cattle, high ground was a berm of brush-covered dunes between beach and marshland. There the cows grazed, chewed cud, and literally ruminated, passing rough forage through a digestive organ, the rumen, that humans lack. Far from appearing panicked, the herd was probably a bucolic sight, from the Greek word *boukolos*, meaning "cowherd."

A close observer, Padilla said, might have noticed subtle differences among the animals: mothers that were watchful or unworried, calves that were playful or lazy, obvious loners or pairs licking or grooming each other. Padilla once spent several months studying cow communication—I found the urge to describe this as "cow-moo-nication" surprisingly strong—by memorizing the free-ranging animals she observed via nicknames like Dark Face and Black Udder. (She didn't realize at the time that the latter was a perfect punning reference to the classic British TV comedy *Blackadder.* What is it about cows and puns?) On Cedar Island, Padilla said, there wasn't simply a herd that was facing a storm. There was a group of individuals, each with its own relationships, including what Padilla doesn't hesitate to call friendships.

Dorian arrived in the purest darkness of the first hours of September 6. Three days prior, it had ravaged the Bahamas with 185-mile-per-hour winds, tying the all-time landfall wind-speed record for an Atlantic cyclone. Some observers suggested giving it a rating of Category 6 on the five-point scale of hurricane strength. It had weakened by the time it reached North Carolina, but it was still a hurricane. Thick clouds blacked out the moon and stars; Cedar Island's scattered lights hardly pierced the rain. Passing just offshore on its way to making true landfall at Cape Hatteras, the hurricane lashed the Pamlico and Core Sounds into froth and spray and sent sheets of sand screaming up the dunes. The scrubby

canopy under which the cows likely took shelter, already permanently bowed by landward breezes, bent and shook in the teeth
of the storm. A 110-mile-per hour gust on Cedar Island was the
strongest measured anywhere in the state during Dorian's passage.

When the eerie calm of Hurricane Dorian's eye passed over the
island, dropping wind speeds to only a strong breeze, there seemed
to be little more to fear. There was still the back half of the storm to
come, but Cedar Island residents, both human and not, had seen
worse. Even in the offseason, the North Carolina shore has hurricanes on its mind. If you see footage of a beach house collapsing in pounding surf, chances are it was shot on the Outer Banks.
Drive around Down East and you'll see many houses raised onto
twelve-foot stilts; in some homes, you reach the first floor by elevator. Maps show that much of the Outer Banks, including most of
Cedar Island and huge swaths of mainland, will be underwater with
a sea-level rise of just over a foot. Residents aren't rushing to leave,
though. A hardened sense of rolling with the punches prevails.

Yet with Dorian, something unusual happened as the center
of the storm moved northward. At around 5:30 a.m., Sherman
Goodwin, owner of Island's Choice, the lone general store and gas
station on Cedar Island, got a call from a friend who lived near
the store. A storm surge was rising in the area, the friend said.
Fifteen minutes later as Goodwin drove through the dim first light
of morning, the water was deep enough to splash over the hood
of his Chevy truck, which was elevated by off-road suspension and
mud-terrain tires. "It came in just like a tidal wave," Goodwin said.
"It came in *fast.*"

By the time Sherman and Velvet, his wife—"My mother really
liked that movie *National Velvet*," she told me—reached their shop,
they had to shelter in the building. Velvet saw a frog blow past a
window in the gale. A turtle washed up to the top of the entryway
stairs. "It came to within one step of getting in the store," Sherman
said, referring to the water. A photograph shows the gas pumps
flooded up to the price tickers.

To understand what happened on Cedar Island that morning,
imagine blowing across the surface of hot soup, how the liquid
ripples and then sloshes against the far side of the bowl. Dorian
did the same thing to the Pamlico Sound, but with a steady, powerful wind that lasted hours.

The hurricane pushed water toward the mainland coast, which in the words of Chris Sherwood, an oceanographer with the US Geological Survey (USGS), is "absolutely perfect" for taking in wind-driven water. The Bay, Neuse, Pamlico, and Pungo Rivers all flow into the Pamlico Sound through wide mouths that inhale water as readily as they exhale it. Much of the rest of the shoreline is an enormous sponge of marshes. What accumulated in this series of reservoirs was, in effect, a pile of water held in place by the wind.

People who know North Carolina's sounds are aware of the tricks fierce wind can play. Coastal historian David Stick once noted that, during a hurricane, half a mile of seafloor in the lee of the Outer Banks can be left exposed as sound water is pushed westward. When that happens, a bizarre phenomenon can occur: A storm surge can come from the *landward* side, striking offshore islands in what's sometimes called sound-side flooding. Scientists know it as a seiche (pronounced *saysh*).

When Dorian's eye passed the Pamlico Sound, the seiche the storm had created began to collapse. Then winds from the southern half of the hurricane, which blow in the opposite direction from the storm's leading edge, drove the water back the way it came. In a sense, the seiche was also running downhill; the ocean tide was falling in the predawn hours, while the hurricane, still pressing down on the Atlantic, forced water eastward, leaving behind a depression. These forces combined to send the seiche pouring out of the Pamlico Sound east toward the Atlantic, nine feet above the water level in the ocean.

The avalanche of seawater was truly vast, equal to about one-third of the average flow of the Amazon River, by far the highest-volume river on earth. The Amazon, however, meets the sea through a gaping river mouth. Dorian's sound-side surge was trying to reach the open Atlantic past what amounted to a levee of Outer Banks islands with just a handful of bottleneck channels between them. At the southern end of the Pamlico Sound, there was an added obstacle: Cedar Island.

The water didn't go around the island. It washed right over it.

The surge left nearly as quickly as it arrived, carrying on to the Outer Banks, where it hit the island of Ocracoke with a wall of water higher than anyone there had ever seen before. Once Dorian passed, floodwaters began receding. On Cedar Island they left thick, greasy muck in buildings and debris on the roads, but no serious injuries were reported. More than a third of the buildings on Ocracoke were damaged, but there were no known deaths.

The first news of losses from Cedar Island's herds of horses and cattle came as soon as the ocean had calmed enough for islanders to go back to sea in their boats. "That's when they saw a lot of them," Styron said. "You know—floating." That Cedar Islanders do not wear their hearts on their sleeves about such things is strongly conveyed by an anonymous source's reaction when I asked how people felt about the dead animals. After an uncomfortable pause, he said, "You can pretty much guess that." Then he added, "Mother Nature allowed them to be here, and I guess Mother Nature can also take them away."

If anyone witnessed what transpired with Cedar Island's feral herds, they haven't said so publicly. Most likely, though, no one saw it, since the surge came without warning in the darkness, and the horses and cows often roamed far from people's homes. The animals would not have been sound asleep in the predawn—feral creatures, like wild ones, are more vigilant through the night than human beings tucked tight in their homes. Still, they may have dropped their guard, sensing that they'd survived another hurricane.

Then suddenly, the sea moved onto the land. Nine feet of water covered the beaches. It drowned the marshes where the cattle fed on sea oats and seagrass, and flowed over the lower dunes. We know from Padilla's research what the scene must have sounded like: high-pitched, staccato mooing—cows' alarm calls—ringing out in the humid air, the bawling of calves competing with the howl of wind and surf. In waters rising at startling speed, mother cattle would have raced to find their young, as bovine friends struggled not to be separated.

Twenty-eight horses were swept away. No one knows exactly how many cows were carried off—four of them managed to remain on land, and locals would later estimate that between fifteen and twenty were taken by the flood. The water likely lifted the animals off their hooves one by one, first the foals and calves, then the adults. They disappeared into the tempest.

The islands known collectively as the Core Banks, located southeast of Cedar Island, are nearly forty miles long and rarely a mile wide. On maps they look like a skeletal finger pointing ruefully toward the North Atlantic. Like most barrier islands they're low—about eight feet above sea level on average, with the highest dunes cresting twenty-five feet—and the whole of them are protected as the Cape Lookout National Seashore. Hurricanes always rough-house barrier islands, but on the morning of September 7, 2019, the day after Hurricane Dorian hit, it was clear that this storm had been a beast of a different order.

Ahead of the cyclone, North and South Core Banks were broken by a single passage, Ophelia Inlet. After the storm, there were 99 additional channels through the islands—the banks had been sliced into 101 pieces. It didn't seem right to call these cut-throughs inlets. They formed as outlets: The seiche that poured over Cedar Island then collided with the barrier islands, and when it did, it bored right through them. "We had never in the collective memory of the park seen a sound-side event like we saw after Hurricane Dorian," said Jeff West, superintendent of Cape Lookout National Seashore. "I did take quite a ribbing about the fact that I lost 20 percent of the park."

West was on the first maintenance boat to sail from Cedar Island for the Outer Banks. Docking at a Park Service site a few miles up North Core Banks, he began driving an ATV along the beach. Fifty feet later he reached the first cut-through and, wading into it up to his neck, found an animal carcass. He didn't take the time to determine whether it was horse or cow. "Sometimes large fish find them tasty," he told me.

Cape Lookout staff would eventually locate the bodies of nearly two dozen dead horses and cattle, along with deer and seabirds. Most were arrayed along the open-ocean side of South Core Banks, likely having passed through Ophelia Inlet before washing up on the beach. The most far-flung horse and cattle carcasses were found near Cape Lookout Lighthouse, about thirty miles from where the animals first washed into the sea.

Cape Lookout workers buried the bodies that the tides didn't take away.

Most of the media coverage of Dorian's aftermath focused on the damage on Ocracoke Island. The first report about Cedar Island's lost herds mentioned only that horses had drowned; the cows had to wait for follow-up articles. It was a blip in the news cycle, soon forgotten as Democrats in Congress sought to impeach Donald Trump.

A pressing question: Can cows swim? Yes, they can. Think of the Wild West, where cowboys guided their herds across deep rivers to fresh pasture or to market. The Cedar Island cattle had been seen swimming, too. One regular visitor described "little bitty calves" lining up to make a crossing to Hog Island, just southeast of Cedar Island in the Core Sound. "I'm like, 'Don't go. You're not gonna make it. It's a quarter-mile swim,'" he said. The calves made the trip with ease.

But it's one thing to cross a narrow channel in calm seas, and quite another to swim through a hurricane. Only the sunniest optimist could have hoped for survivors from Cedar Island's herds. "I'm thinking the way the wind was blowing, it was extremely hard to keep your head above water, swimming when you have waves crashing over," said Pam Flynn, a retired kindergarten teacher and a Down Easter since 1972, who went looking for surviving animals. "I feel like their last few moments were torture and pain and fear. It was heartbreaking."

A month passed. Wind and waves quickly filled in the channels created by the storm, but what was formerly the southern end of North Core Banks lingered on as a separate island: Middle Core Banks, which would stand alone for two years. One day in early October, members of a Cape Lookout resource-management team hopped on their all-terrain vehicles for a routine sweep up Middle Core Banks—almost daily, they'd search for sea turtle and bird nests in need of protection from the fond American pastime of driving on beaches. This time they spotted something else: the tracks of some large animal or other. They were too big to belong to a deer, and, with two toes instead of a hoof, could not have been made by a horse. They had to be the prints of a cow. A Cedar Island cow.

"Initially," West said of being informed about the prints, "I did not believe it."

Then the resource team sent him photos of the tracks, and West knew he had to see this survivor cow with his own eyes.

West grew up on a ranch near Temple, Texas, and had experience tracking cattle. It seemed like he might need it. In the days after the prints were discovered, the cow that left them proved elusive; to West's knowledge, no one from the National Park Service had yet seen it. Cedar Island cattle are often active at night, moving swiftly like pale apparitions, and although Middle and North Core Banks are so narrow in spots that you can walk from the sound side to the open Atlantic in three minutes, much of the land is a labyrinth of ponds, marshes, and fly-infested thickets. Additionally, resource crews had spotted hoofprints on small adjacent islands—despite the recent seagoing drama, it appeared that the cow was now making short water crossings too. "No fear of swimming, none at all," West said, with admiration in his voice.

In the end, he found the animal by accident. West had taken a boat out to Long Point on North Core Banks, home to a cluster of rustic wooden cabins that, in more ordinary times, the Park Service rented to visitors. Dorian's storm surge had razed two heavily fortified structures that provided electricity and treated water to the wind-battered huts. And there it stood, chewing grass—a dune-colored cow among the dunes, with a coat like gold sand blown onto white sand. It was well muscled, a little heavy, basically an ordinary cow.

"'I'll be damned. There is a cow here,'" West recalled saying aloud. "Nothing like your own eyes seeing it."

At the sight of West, the cow's eyes got big. Then it ran away.

West knew that he would need to relocate the cow, both for its own sake and to preserve the wild habitat of the park. For the moment, though, the Cape Lookout staff were too busy assessing and repairing Dorian's damage to deal with a wayward bovine. Meanwhile, rumors of the survivor began to trickle out as visitors returned to the Core Banks and saw tracks. Pam Flynn and her friend Mike Carroll were among them. "We kept going back and back," said Flynn, until they lucked into a sighting. "We were so excited to see those cows."

Not one cow, then, but *cows*: three in all. There was the classic bleached-blonde that West had seen; another one with large, light brown spots, like a map of the ancient continents; and a pale young adult, possibly the spotted cow's calf. Somehow they had survived, found each other, and formed a compact herd. "It just renewed my faith that there are good things in life, something at the end of the rainbow," Flynn said. "You know, a little sign that we'll be OK, we'll get through this and go on."

On November 12, the *Charlotte Observer* broke the story of the survivor cows, and a media circus ensued on Cedar Island. One unfortunate local figure, wrongly described in the press as the cattle's owner or caretaker (they have neither), had reporters knocking on his doors and chasing him up his driveway. On television especially, the tale of survival was presented as a quirky good-news story. The *Virginian-Pilot* would go on to call the cows "the cattle that enraptured a nation."

The hook of the story was its element of surprise: We see cows as stupid, physically awkward, mildly comical brutes, not heroic fighters. The media made heavy use of puns, of course, giving the life-and-death story a chuckling, children's-book quality. Hurricane Dorian had come ashore "like a cattle rustler in the night" and "corralled" the animals. The cows' survival was an "udder miracle." An awestruck Raleigh *News and Observer* tweeted, "Four miles on the moooooove? Who knew cows could swim that well?"

To estimate how far the cows had paddled during their ordeal, journalists seemed to have measured the shortest distance between Cedar Island and the Core Banks using digital tools like Google Maps. Most put the swim at four miles; NBC preferred the precision of 3.39 miles. But when Alfredo Aretxabaleta, an oceanographer working with the USGS, saw one of these straight-line measures, he spied a problem. "During a storm, I just don't think that's the path they would take," Aretxabaleta said. He suspected their journey was longer—much longer.

Aretxabaleta studies the trajectories of objects adrift, using computer models of wind, tides, and currents. He sometimes throws trackable equipment into the sea to float where it will; the science has been jokingly called driftology, but it has repercussions for our understanding of how climate change could affect coastal erosion, where oil spills and other contaminants might flow, and where to carry out maritime search and rescue work. "In a way," Aretxabaleta said, "the case of the cows is a kind of search and rescue."

Coincidentally, Aretxabaleta grew up in Spain's Basque Country, on a farm where the cattle took dips in an irrigation pond. (His assessment: "They are *not* good swimmers.") After Hurricane Dorian, Aretxabaleta in his spare time began to model the probable trajectory of the Cedar Island survivor cows once they were swept out to sea. What emerged was far different from the image of cows taking the shortest route across the Core Sound.

In the context of Aretxabaleta's model, the sea, in the gray pall of first light as the cows are carried away, is a chaos of riptides, breakers, and blowing spray. With the cows' eyes only inches above water, land is quickly lost from sight among swells as high as ten feet; from the perspective of a single cow, it's nearly impossible to keep eyes on the rest of the bobbing herd. Each is fighting not so much to swim as to remain afloat. The currents and tides, made stronger by the force of the storm, are in charge.

The animals are first pushed rapidly southeast along the coast of Cedar Island, then into the center of the Core Sound, where they're gradually drawn close to the powerful outflow at Ophelia Inlet. But as the tide changes from ebb to flood, Ophelia no longer sucks the animals toward it, but pushes them away. With the ocean now flowing into the sound, the herd are swept back to the north. At last the tide switches again, and Core Sound has many dozen new channels through which to send water back to the Atlantic. Like in a tub with many holes, though, it's the large ones that have the most pull. Any animals still alive are drawn again toward Ophelia Inlet.

The prospect of passing through any channel would be a fearful one. Surfers sometimes dig cut-throughs between the sea and fresh water that has pooled behind dunes; the flow generated in such canals can resemble a river rapid, with waves large enough to surf. The Core Sound is not much calmer. After the cattle are washed off Cedar Island, the wind doesn't drop below gale-force for seven hours, and whitecapped waves linger much longer. Though the Core Sound has shallow areas such as sandbars, Aretxabaleta accounted for them in his simulations and says it's unlikely that any cow found footing for long, if at all, during its journey.

His model explains how the cows and horses that were found dead on South Core Banks ended up where they did, flushed through Ophelia Inlet and then strewn to the south by the open Atlantic. By his estimation, none of the survivor cows swam four miles on a straight-line path. In fact, Aretxabaleta said, the probable routes taken by the cows, whether living or dead, range from 28.5 to nearly 40 miles. At the low end, that's considerably greater than the distance across the English Channel. It's more than ten times what swimmers complete in an Ironman triathlon. By Aretxabaleta's measure, the absolute shortest period a cow would have been in the water is 7.5 hours; the longest is 25 hours.

"If it had been humans, it would have been incredible—I mean, like Robinson Crusoe," he said. "The fact that those three cows survived is something close to a miracle."

*

Suppose we didn't settle for miracles, much less the "udderly miraculous." Suppose we refused to consign the three cows' survival to fate and chance. There are other factors we might consider, each of which drifts toward reckonings with how humans interact with bovines.

The first possibility is that the Cedar Island cows were able to endure their ordeal because they were a breed apart, not metaphorically but literally. Blood type and DNA tests suggest the feral horses that live on Cedar Island are likely descendants of Spanish colonial horses, which first came ashore in the United States with Juan Ponce de León in 1521. The cows may have Spanish colonial blood too; no one knows, though, because their genetic makeup has yet to be studied. What's certain is that cattle have been abandoned or shipwrecked along North Carolina's coastline since at least 1584. The Cedar Island cattle could have more than four centuries of heritage.

Spanish colonial cattle are different from the commercial breeds that predominate today. "They're long-lived, they're good mothers, they'll eat things other cattle won't," said Jeannette Beranger, senior program manager at the Livestock Conservancy in Pittsboro, North Carolina. "And they're smart. The locals will tell you, 'Be careful. They'll eat your lunch!'"

They are also notoriously tough. In the days before the Civil War, Spanish-descended Pineywoods cattle, for example, were known for heat tolerance, disease resistance, and a capacity to live in landscapes too harsh for commercial breeds. The rugged nature of the Pineywoods cows resulted in a markedly different relationship between them and their owners than we see in today's industrial agriculture. Some ranchers had so much respect for their cattle that they would not tolerate the use of dogs to harass the animals during roundup. Others felt it unfair and demeaning to confine the cows with fences. It was only in the 1950s, with commercial feed and motorized equipment used to clear and mow pastures, that the Pineywoods herds began to fade, though a small number of farmers in the Deep South breed them to this day.

Phillip Sponenberg, a veterinary scientist who has spent fifty years searching for the purest-blood remnants of Spanish livestock in the United States, sees signs that the Cedar Island cows share at least a trace of that ancestry. "Some of them are basically white, but they have dark ears, eyes, noses, and feet. That's a fairly unique color pattern and, in North America, often of Spanish origin," he said. Some of the Cedar Island cattle also have horns that twist like a Spanish colonial cow's.

Several experts I spoke to suggested that the fact that any cows at all survived the Dorian surge is clear evidence that they aren't ordinary cattle. Most agreed that no modern breed would have made it through such a disaster. In this there is recognition of how we've degraded cattle as animals, turning them weak and needy. It also feels too convenient. It allows us to duck a more uncomfortable possibility, which is that these animals that most of us readily eat may have made it through the storm by drawing on the same internal resources that humans do in extreme circumstances. Not just a hardwired survival instinct, that is, but a fierce desire to live—one strong enough to sustain hour upon hour of mortal struggle.

I should pause here to say that I eat beef. I put cows' milk on my cereal. I have leather shoes and belts in my wardrobe. Still, like many other people, I recognize that rearing and slaughtering cattle raises issues that are ethically complicated, contradictory, and sometimes deeply weird. None of this, however, is what led me into the terrain of cow psychology. Instead, I simply wanted to know why one cow might survive swimming through a hurricane while another might not.

Remarkably for an animal domesticated thousands of years before the dawn of civilization, the scientific study of cows distinct from their roles as livestock is mostly a recent pursuit. When Mónica Padilla de la Torre reviewed existing research on cow communication more than a decade ago, she was surprised to discover that almost nothing had been done on the subject—which is why she started from scratch, watching cattle through field binoculars like a Dian Fossey of the rangelands. "I think we have a moral responsibility to know these animals that we have lived with for so long," she said.

For a 2017 paper, Lori Marino, a biopsychologist, reviewed every study she could find on cow psychology. Again, the trove was not impressive. "There's a lot to learn about these animals," said Marino. "There is resistance to coming to terms with who they actually are, their cognitive and social and emotional complexities."

The problem, of course, is that those complexities could upend our relationship with the species. Marino describes the prevailing way we think about cows as an ideology, one that frames them as dull creatures that are fine with their lot in life, even if that life includes crowding, untreated lameness, being burned with a red-hot iron, and having their calves taken away—practices common in modern industrial farming.

In Marino's review of the available research, however, she found that cows are "very sensitive to touch," and that they respond to injury or the threat of it in ways similar to dogs, cats, and humans: by avoiding causes of pain, by limping, groaning, and grinding their teeth, and by evidencing higher levels of stress hormones in their blood. On the other hand, pain and stress, and especially their severity, may be more challenging to recognize in cows, since they evolved to avoid showing signs of weakness, which can attract predators. Cows are stoics; they tough it out.

Though data on cow psychology is limited, I still found it surprising. It was somehow troubling to learn that cows readily recognize one another and are able to distinguish cattle of any breed from other sorts of animals. Cattle are able to navigate and memorize physical mazes with flying colors, outperforming hens, rats, and even cats, and leading researchers to conclude in the study that "the problems were too simple." When cows were tested in more complex mazes, one in five succeeded at the toughest challenges, and could recall how to navigate the maze when retested six weeks later.

Here we enter territory more meaningful to the question of how those three cows might have survived swimming through a hurricane, since mastering mazes involves not just intelligence but also motivation. It's true that only one in five cows solved the difficult mazes, but that may be because they dislike being alone and are fearful of places with many potential hiding places for predators, such as a maze. Throughout the tests, some of the cattle, despite a food reward for completion, appeared to resist, give up, or become fearful. Others were bolder and more curious. "This may," the researchers reported, "suggest the possibility of the involvement of personality."

With cows, some of the clearest expressions of apparent personal motivation are found in near-death escapes from slaughterhouses. In one of the most famous examples, a 1,050-pound cow broke loose from a Cincinnati facility in 2002. After jumping a six-foot fence, the cream-colored bovine was seen on a nearby side street, was subsequently spotted on a major parkway, then finally escaped into a wooded city park. Over the next eleven days, it evaded the SPCA, traps, tranquilizer darts, even thermal imaging from a police helicopter, before finally being captured.

The animals we eat are nameless, yet escaped cattle that make the news are often rewarded with names. Once that happens, they are unlikely to be returned to industrial production. In this instance, the cow was dubbed Cincinnati Freedom, and lived out her days at a rescue shelter where she was standoffish with people but bonded with three other slaughterhouse escapes. When "Cinci" was dying in 2008, her cohorts attacked the car of an attending veterinarian.

The prevailing ideology, to borrow Marino's term, has been to explain away cattle's responses to the world around them as exclusively innate or instinctive. By this standard, when the herd of cows was swept off Cedar Island into a violent ocean, survival would have been determined by luck and physical strength.

If individual cows have personalities, perhaps not as complex as our own, but no less singular, then that assessment may need to change. Once the storm had washed the herd into the ocean, some of the cattle, stricken by panic, would have quickly succumbed to water inhalation or exhaustion. Others, dragged farther and farther from land by the powerful currents of the seiche, might gradually have lost the spirit to fight on. But is it conceivable that three would keep going, drawing on exceptional mental toughness to push their bodies far beyond anything they'd endured before, in order to survive?

"I would use 'willpower,'" Marino said. "I wouldn't hesitate to use that term."

No one will ever be certain exactly what the cows went through. Did the two that were later seen ashore together also make the swim that way? We don't know. But we can hypothesize that the cows in the water would have *tried* to stay together. Studies show that even being able to see another cow reduces their stress. Together, they may have faced calamity with less fear. Perhaps that alone made the difference.

We can picture the three cows desperately blinking their eyes against the waves and the wind-driven spray, enduring the creeping cold in their bodies, the gradual ache and depletion in their muscles, the thirst and hunger after what may have been hours at sea, the maddening whine of the wind. Then finally seeing, or perhaps first smelling, land once again. Hearing the roar of the fearsome inlets and fighting to avoid being sucked into one.

Their hooves making contact with the sand.

Scrabbling to gain footing.

Surging onto the land as the water rushed between their legs, then dragged back toward the violent ocean.

Finally walking free, with a feeling like profound relief to be alive.

The question of what happened next can perhaps be told through another tale of animal survival. When Hurricane Fran struck in 1996, the storm surge that hit New Bern, North Carolina, flooded the offices of an auto salvage business to a depth of sixteen inches. Inside was a junkyard dog named Petey, who stood ten inches tall. After the flood retreated, Petey's owner found his dog alive but exhausted. When he saw that Petey was soaked with muddy, oily water up to its neck, he surmised that his pet had dog-paddled inside the building for as long as eight hours to survive. Here's what animals do after such an ordeal: Petey slept for two days straight.

Though little used this way today, we do have a word for bovines that roam free like mustangs. They are *mavericks*. The term has roots in one Samuel A. Maverick of Texas, whose unbranded cattle got loose into the landscape around 1850. In one version of the story, the force that scattered his cows was a hurricane.

It's fitting, then, that on November 21, 2019, it was the duty of six cowhands—complete with lassos, chaps, and spurs—to track down the three mavericks on North Core Banks. One of the men carried a rifle loaded with tranquilizer darts and Jeff West drove a Park Service ATV next to the cowhands astride their horses. The plan had always been to get the cows home, said West. That fact had not prevented fierce debate from breaking out online.

"Some people thought we should just kill them, be done with it," West said. "Some people complained, 'Why are we spending taxpayer dollars on this?' Heard that more than once. Some people said we ought to just leave them alone, let them exist out there on the banks."

Many assumed that the cows had survived only to be sent back to owners who would fatten them for slaughter. On the Cape Lookout National Seashore's Facebook page, a theme emerged that the cows deserved to live; through baptism by flood, they had transcended their place in the scheme of things. "If they have to be removed then take them to a sanctuary. They deserve life. Do not turn those babies into meat after what they've survived!" wrote Misty Romano. Don Riggs of Asbury, New Jersey, wrote, "Really? Why not just bypass the farm and go straight to the slaughter-house?" Judy Cook of Oak Island, North Carolina, simply declared the cows "as cool as the horses."

Modern views about cows are messy. Many of us, if not most, seem capable of holding somewhere in our heads the idea that cows are sentient beings that we should have compassion for, but also of suppressing that idea enough that we allow them to suffer cruel conditions along the way to being killed for our benefit. Jessica Due, senior director of rescue and animal care for Farm Sanctuary, an organization devoted to ending the agricultural exploitation of livestock, tells a story that exemplifies the ways this can play out. The sanctuary has been called more than once by the same man to come and rescue an animal from a slaughterhouse. The man is the owner of the slaughterhouse. He calls on those rare occasions when a cow gives birth while being processed. This is where he draws the line; he strongly prefers not to kill these mother cows. Otherwise, he oversees the deaths of cattle on an almost daily basis.

Curiously, just as research is emerging in support of the idea that cows are something more than most of us thought they were, they are also under scrutiny as environmental polluters. Cattle are blamed for producing 9 percent of global greenhouse gas emissions, including their famously methane-heavy belching and flatulence. Cows swimming in a hurricane: It could be a Hokusai print for our times. As a result, progressives and vegans look forward to a future with far fewer cows—to save the planet, to protect the animals from our cruelty, or both at once. Many in the industrial beef industry, meanwhile, remain reluctant even to concede that cows are meaningfully sentient. In the ten thousand years of human-cow relations, it's possible that cattle have never had as few supporters as they have today.

Stephen Broadwell, the leader of the cowhands trotting down North Core Banks nearly three months after Hurricane Dorian, is one of those supporters. Broadwell is russet tanned and often wears a cowboy hat, but that is where the stereotypes end. He was raised in corn, tobacco, and soybean country, where North Carolina's Piedmont Plateau meets the Coastal Plain. Yet he dreamed of being a rancher. "It's one of those things—I guess it's born into you," he said. At the age of thirteen, he took a summer job on an 80,000-acre ranch in southern Colorado, and that was that. He was a cowboy.

After graduating early from high school, he earned a veterinary assistant's degree and was soon hired on at 3R Ranch Outfitters in the foothills of the Wet Mountains southwest of Pueblo. It was his immersion in an approach to ranching that attempts to mimic natural systems. "Our neighbors were thinking that we had this magical paradise for a ranch around them, and it was just the management practices they'd put in place years ago," Broadwell told me. "That really got my motor going."

The company he runs today, Ranch Solutions, might best be described as a holistic ranching consultancy. Broadwell will come to your property and do pretty much anything you need, including building a house from scratch and putting your first cows out to pasture. He has one rule, however: He will not help you raise more cattle than your land can sustain. He has photos of his team riding through the lush, knee-high grass of a client's property. It's a field that had already been grazed, but with the cattle moved off before it was eaten to the ground. The pasture was fertilized by manure and supplemented by cover crops that rebuilt nitrogen in the soil during winter, leading to grassland that sequestered more carbon. A cattle ranch, as Broadwell would have it, is an ecosystem.

The claim that holistic management can achieve this state is hotly contested, but research has lately suggested that yes, cattle can live and die without contributing to climate change. (And let it be noted that there is a strong pot-calling-out-the-kettle factor here, given that the average American human's carbon footprint is twice that of the average American cow's.) But we need to raise fewer of them, graze them in ways that mimic natural systems, and keep them off land better suited to food crops.

The future of cattle farming, in other words, may look a lot like the Cedar Island herd. Here are cows that can survive heat that would wither modern breeds, in a landscape where nothing we farm will grow. Here are cows adapted to eat what almost nothing else can. "It's what a billy goat would *not* want to eat," Broadwell said. Here are cows that are disease resistant, drink brackish water, defend themselves from predators, and generally require very little in the way of carbon-intensive coochie-cooing. They are the kind of cows that in the past demanded our respect, and one day might again.

"I grew up with stories from my older relatives about working cows in the river breaks"—steep cliff and canyon country—"and how they were more like deer than cows," said Jeff West, remembering his youth in Texas. "We ran some cows out in North Fort Hood military reservation, and we only messed with them one time of year, during the roundup. Some of those cows were pretty feisty. But not like these Cedar Island cows. I've never run across any cows like these cows."

When Ranch Solutions and West arrived on North Core Banks for the roundup, they had a plan to haze the survivor cows out of the marsh grass, which grows in muck that's sometimes deep enough to swallow a horse to its belly. Then there was the chaparral. "*Thick* is a poor word to describe it," West said. "It is intolerable of somebody passing through." It took a long time to locate the cows, and then to work them out into the open so that each could be shot with a dart. Sedated, two of the three became pliant enough to be led back to a trailer that had been ferried to the island.

The final cow, the first to be found after the hurricane—alone—did not become pliant. She fled north, managing to hole up in especially dense and convoluted terrain. The team could just see where she was hiding, and managed to hit her with another dart. Then they waited, sure she would gradually go to sleep. She did not. At last the cowhands tried approaching her.

"And she took off," West said.

Just up the coast was the Long Point camp where West had first spotted the cow a few weeks after the storm. The buildings still stood empty. Wind sucked and blew between weathered wooden walls. Screen doors creaked on rusty hinges. Hooves squeaked in the sand. It was in every way like the setting for a Spaghetti Western shootout. When one of the riders saw a clean line of fire, the crack from his gun echoed among the shacks, then faded into the roar of the tumbling surf.

With three darts' worth of sedation flooding her system and

blood trickling down her pale coat, the cow somehow ran again. She ran out of the camp. She ran up the beach. After half a mile, she couldn't run anymore. Then she walked. "It was O.J. Simpson all over again. It was the slow-speed chase," West said. "It was me and all the cowboys at a walking pace, going along until that cow stopped."

When she finally did, she stared them down. "Like, 'Try me,'" West said. The cowhands closed in, and one last time she managed to run. Then they got ropes on her and brought her down.

From there the job got easier. With the sun on the horizon, they worked a tarp under her prone body and sledded her down the beach. She came to while walled in by the trailer, her fellow survivor cows beside her. Given hay and fresh water, all three refused it.

The next morning, Ranch Solutions ferried the cows back across the Core Sound, drove to Cedar Island's northern cape, and backed onto the beach. It was Broadwell who did the honors of swinging open the trailer's gate. The cows stared at the sudden possibility of escape. They took cautious steps toward the opening. Then they burst from their confines. They ran—*galloped*—down the sand. Heads up, ears forward, they seemed instantly to sense that they were home and free.

On Cedar Island, the return of the cattle brought a sense of normalcy. When I asked one shopkeeper how islanders felt about the cattle now, she responded instantly. "Fiercely protective," she said. No one I spoke to on Cedar Island knew of anyone who'd witnessed the three cows' reunion with the remaining herd—the four animals that hadn't been swept away by the storm in the first place. But according to Padilla, it likely involved muzzling, low and gentle moos, and gamboling. It might also, finally, have involved grief.

People who've looked closely at this issue, such as Barbara J. King, an emeritus professor of anthropology at the College of William & Mary and the author of *How Animals Grieve*, think the blow would have struck hardest when the survivors came home to find the herd decimated. They might have searched the range for missing herd mates and bellowed in an effort to make contact. King, choosing her words carefully, said, "The potential is incredibly strong for the awareness of loss and feeling of distress that would meet my criteria for grief."

Yet home also brought a different kind of surprise. The cow that had fought so hard to avoid capture by the cowhands turned out to be pregnant. Could that have played a role in her survival? If a cow has a will to fight for its life, might it also fight for the life of its unborn calf? "Biologically, it wouldn't be strange to assume that," Padilla said. "She wants the calf to survive."

Two months after being returned to Cedar Island, the pregnant cow gave birth to a healthy calf, as blond as the dunes. It was born, as if to mark what it went through in utero, with one brown eye and one blue. The calf was not given a name, but the mother was: Dori. The name is not an allusion to the character in *Finding Nemo* who sings of how, in hard times, we must keep swimming, swimming, swimming. No: She is named after Hurricane Dorian.

JESSICA CAMILLE AGUIRRE

Another Green World

FROM *Harper's Magazine*

AFTER THE BANG, much later, after our sun took shape and the earth began to emerge from a molten conflagration that hummed and spat and heaved forth gases that would envelop the entire planet; after bacteria in its waters exhaled enough oxygen to encourage inchoate life, an infinitesimally small portion of which would endure billions of years of evolution to plant seeds and build cities and poison the waters from which it came and turn verdant rainforests to toxic dust; and after that life then began to dream of venturing out into the stars, one man stood in a greenhouse near Tucson, Arizona, and thought about how to re-create it all.

The man, an amateur scientist named Kai Staats, went outside and hosed himself down in the 105-degree heat. Clouds were piling up to the east, but Staats was on deadline and couldn't let the prospect of foul weather upset his experiment. Staats was a space man. Although he didn't work for a space agency or have a degree in astrophysics, he had joined a number of endeavors that aspired to launch mankind farther into the cosmos. He had volunteered for a high-altitude ballooning experiment led by an archaeology professor who made DIY space suits. He had worked on a proto-type space base in Utah, and tried to advise a doomed attempt to build a colony on Mars. In Arizona, he was embarking on his most quixotic project yet: the construction of a sealed human habitat, one in which people could exist in a miniature environment unconnected to the earth's atmosphere. Perfecting this technology—known as a closed loop—could allow humans to survive far from the planet for months, enabling long-distance space travel and, eventually, the colonization of other celestial bodies. Staats named his project Space Analog for the Moon and Mars. He called it SAM for short.

SAM wasn't being built from scratch. Instead, Staats was refurbishing an old test module that had been used for a similar project aborted some three decades earlier. Down a dirt road was the site of that experiment: a glass building of serried arches and a stepped pyramid. Biosphere 2, as it was called (Biosphere 1 was Earth), looked like an extraterrestrial cathedral. It consisted of seven biomes—including a rainforest, a marsh, a desert, a farm, and an ocean with a coral reef—stretching across more than three acres enclosed under white domes. In the early nineties, eight people had lived inside for two years, shut off from the world except for sunlight and electricity. They were the first of fifty planned crews; the mission was intended to last one hundred years. Instead, it came to be dismissed as a failure by much of the scientific establishment after nineteen of its twenty-five vertebrate species went extinct and carbon dioxide levels shot up so high that half of the crew developed sleep apnea. Its financier instigated an armed takeover, and now the site serves as a research center and tourist attraction.

As the clouds darkened on the horizon, Staats scrambled to patch up SAM's final few seams. In less than twenty-four hours, he was going to shut SAM's heavy submarine door with a few volunteers inside. The trial was the first step in Staats's plan to build a tiny, exportable world. It would last only a few hours, but it was not without risk. The carbon scrubber, a long silver box filled with mineral zeolites that adsorb carbon dioxide, could fail, causing gas levels to rise to dizzying heights, and compromising the cognitive processing of those locked inside. Or, if the pressure rose too high, the building could explode.

Dripping wet, Staats grabbed some tools and climbed atop SAM's pressure controller. "Hey, Trent?" he called. Trent Tresch, an outdoor expedition guide, was taking time off during the pandemic to help refurbish the test module. "We've got a big storm coming in here."

"We were just talking about that," Tresch replied.

"I'm wondering if we should try to maybe break from our schedule to try and seal that," Staats said, running his finger along a joint. "Maybe I should try and spray-foam those now."

"Okay. We could finish siliconing them too." A goopy white silicone sealant called Dow Corning 795 had filled every crack in Biosphere 2's white tetrahedral frame back in the eighties, so Staats had ordered some online for $14 a tube.

"Well," said Staats. "That may not be a bad idea."

The human body is particular about what it needs to stay alive. Surviving well above or below the earth's crust requires an artificial life-support system to manage changes in pressure and temperature. The body must have near-constant access to oxygen and periodic access to food and water, which means it must also manage the consequences of that consumption: excrement, urine, and carbon dioxide. This isn't much of a problem on brief space flights—the first astronauts who went into orbit wore diapers. But leaving the planet for more than a few months, let alone trying to colonize a new planet, depends on establishing livable conditions in a continuously regenerating way—on re-creating, more or less, the functions of the earth's ecosystems.

Most scientific progress to date has concerned the recycling of abiotic factors: air and water. On the International Space Station, for instance, chemical molecules are arranged and rearranged in a never-ending atomic dance. Astronauts breathe air similar to Earth's, composed mostly of nitrogen and oxygen. The nitrogen is shipped to the station in tanks, but much of the oxygen is produced on board through electrolysis—by running an electrical current through water to divide oxygen atoms from hydrogen atoms. The leftover hydrogen is then transferred into a cylinder called a Sabatier reactor, along with carbon dioxide and a nickel catalyst. The reactor produces methane, which is discarded overboard, and water, which is recycled for drinking or reused to create yet more oxygen. Carbon exhalations that aren't fed into the Sabatier reactor are captured by zeolites, which, when heated, release carbon dioxide into the vacuum of space.

What scientists have not managed to figure out is how to sustainably generate and dispose of the biotic variables of ecosystems, such as food and waste. Without a way to cultivate plants, for example, the International Space Station relies on food shipments that are delivered every few months at a cost of $2,720 per kilogram. (The cost was roughly $18,500 per kilogram until SpaceX's reusable rockets lowered expenses.) The weight of the food needed for a long trip without the possibility of resupply—to, say, Mars—is prohibitive. As for excrement, it's currently suctioned into plastic bags and jettisoned along with other waste onto a cargo ship that incinerates upon reentering Earth's atmosphere. Its fiery streak is often mistaken for a shooting star.

Theoretically, astronauts could grow their own food and return their own waste to the soil. Integrating plants and other organisms into a life-support system, at the right scale, could also make use of their other contributions to the biogeochemical cycle, replacing the technology that simulates abiotic regeneration with living ecosystems. After all, plants and animals, including humans, are constantly exchanging carbon through photosynthesis and respiration. One's effluence is the other's fuel. By the same token, water evaporates, condenses, and returns to the soil, where it's purified. Nature, in other words, is a life-support system that wastes nothing, that requires no resupply, and that humans already belong to, which is why scientists have long tried to re-create it for long-duration space missions.

But using biological systems for space exploration always poses the risk of losing control. When you're dealing with plants, soil, or microbes, things happen beyond human command, even in a highly monitored environment. An aphid can lay waste to the tomatoes, or bacteria can infect the carrots. In big, complex ecological systems, one small imbalance won't destabilize the whole, but in simpler systems, any slight mismatch in power dynamics can allow one organism to gain a foothold and take over. Living systems thrive in their complexity, whereas artificial life-support systems need to be manageable, reliable, and obedient.

Before starting SAM, Staats tried to account for biotic complexity by creating a computer model of a regenerative life-support system. This was how I first heard about him. I was looking into the history of Biosphere 2, and learned that he had gathered data for his model at the original site. In early 2020, I called Staats to ask about his research, and he mentioned that he was going to try to restart a scaled-down version of the experiment, but with the explicit goal of modeling a Martian colony. He sent me a project brief that included plans to relocate the entire test module to the lawn in front of Biosphere 2's original entrance. It was an ambitious idea. The module would be expanded with shipping containers, and the entire structure would be pressurized and

hermetically sealed. A few months later, he invited me to join the
crew that would lock itself inside the rehabilitated test module for
its first test run. I had never felt the pull of Mars, but I was curious
about what our attempts to go to outer space could tell us about
living on this planet. I said yes.

Biosphere 2 sits on a sprawling campus at the edge of a town called
Oracle. When I arrived at the gift shop last June, Staats was waiting in the shade, a pair of wraparound sunglasses perched on his
baseball cap and a T-shirt tucked into his utility pants. The son of
a Lutheran pastor who preferred fixer-uppers, Staats could wield a
table saw by age ten. He studied industrial design at Arizona State
University, then built a software company, Terra Soft Solutions,
which developed an open-source operating system called Yellow
Dog Linux. In 2008, he sold the company and made a series of
documentaries about the Palestinian territories, the South African
Astronomical Observatory, and LIGO, the gravitational wave observatory. On the side, he helped to design a playground in Poland
and to build the first astronomical observatory in East Africa. Staats
can't remember the last time he had a regular paycheck. When the
pandemic started, another documentary project fell through, and
he made ends meet by fixing cow fences.

All the while, Staats nurtured a growing interest in space travel.
In 2012, he received an invitation to join a crew at the Mars Desert
Research Station—one of two analogue space missions run since
the early aughts by the Mars Society, a nonprofit. He spent two
weeks living with a group of scientists and engineers near Hanks-
ville, Utah, in a model lander-cum-habitat. He also worked with
the founder of Mars One, a Dutch company aspiring to build a
permanent colony on the red planet. In 2019, after collecting millions of dollars in fundraising, Mars One declared bankruptcy and
was derided as a scam. But its demise didn't deter Staats so much
as confirm his suspicion that the old model of space exploration,
dominated by state-run agencies, wasn't as unassailable as it once
was. With enough vision, patience, and money, he thought, just
about anyone could contribute in some way to helping the species survive off-planet. Bas Lansdorp, the founder of Mars One,
may have been a con artist, but that didn't mean that Elon Musk,
Jeff Bezos, and Richard Branson would all end up in bankruptcy
court. The field's fragmentation was an opportunity, a chance for

closet engineers like Staats, with access to YouTube and NASA's published papers, to build backyard rockets and develop their own space suits.

According to Staats, the human expansionist project is running out of time. "There really is a limited number of resources," he told me, referring to the fossil fuels currently needed to power rockets. "If we don't use what resources are left now to become interplanetary, we will have lost the window and never get off the planet." ("The alternative to that, of course, is the space elevator," he added, referring to a futuristic engineering proposal that involves fixing a cable from Earth to a satellite at least 22,000 miles away.) This is a concern he feels viscerally; the thought of being stuck on Earth makes his chest tighten. While he doesn't consider himself a pessimist, Staats is increasingly certain that human civilization is on a path to self-destruction. Space colonization, as he sees it, is our only option.

Staats walked me through the desert brush, past some abandoned buildings to the test module. He showed me an empty shipping container that would eventually become living quarters, as well as a greenhouse, pocked with holes, in which he planned to build a replica of a Martian crater. "We have a lava tube that's going in that corner," he said. "So we can go to the top and actually rappel into the lava tube with space suits." A company that made synthetic rocks for zoos would build the landscape, and there were plans to install a movie stunt harness to replicate low gravity.

The module itself was a small trapezoidal structure made of steel and glass. Behind it was the lung, a giant white disk designed to calibrate pressure with a large metal pan that would rise or fall as changes in temperature caused the air inside to expand or contract. The exterior wall was marked n1987b. When I asked Staats what it meant, he told me he had registered the module as an experimental aircraft with the FAA. The stencil was its tail number.

Inside, a metal frame made of interlocked tetrahedrons rose up about twenty feet. In an attempt to re-create conditions on Mars, which is fifty million miles farther away from the sun than Earth is, Staats had tinted the windows and painted over the roof. The floor sloped down gently to a basin in the center, and in the back, an underground tunnel led to the lung. There was an electrical panel by the door, an air-conditioning unit near the rafters, and a spigot. Other than that, it was empty, save for a shelf that would later hold the carbon scrubber. "One day we walked by this dilapidated building, just overgrown with cactus and rattlesnakes, and when I looked in the lung I nearly puked, it was so nasty," Staats said, recalling the first time he saw the test module. "I said, what is this amazing building? I could see the potential. It looks like a spaceship."

On the day of our enclosure, a cavalcade of trucks and vans arrived carrying students from the University of Arizona's Controlled Environment Agriculture Center. They were going to install racks of lettuce and spinach to evoke the spirit of the original Biosphere 2 experiments. The students unloaded a few metal shelves and began setting up fluorescent lighting. The roots of the plants, entwined in spongy mineral wool soaked in fertilizer, were loaded into trays and long PVC pipes, rigged up with pumps that would continuously circulate water. Staats radioed John Adams, the Biosphere 2 deputy director: the students needed glue. By midafternoon, a few hours behind schedule, hundreds of lettuce and spinach plants tufted haphazardly from the pipes.

After lunch, Tresch appeared with the carbon scrubber and began filling it with zeolites. He had finished assembling the machine on the kitchen table in the apartment he shared with Staats. Adams, and Katie Morgan, one of Staats's colleagues, arrived as well; they would be the other two volunteers with us inside SAM. Morgan had brought a bottle of champagne to celebrate if the module managed to hold its seal. She reminded everyone that no matter what happened, she had to be on a flight to Jacksonville, Florida, in a few days because her family was participating in a golf cart decorating competition for the Fourth of July. They would all be dressing as alligators. "If you die, I'll fly out," offered Morgan's friend Britney, another Biosphere 2 employee.

"Let's walk through procedures and establish some goals," said Staats. He didn't think we would have to be enclosed for more than four hours before we would start to see the kind of impact he expected: a rise in carbon dioxide and a decrease in oxygen as we depleted what had been sealed inside. "John, you expressed some concern about the oxygen being too low. What point do you think is too low?"

"I don't think we want to get below 15, 14.5 percent," Adams said.

Oxygen is only about one fifth of the air we breathe; the rest is mostly nitrogen, with a small amount of argon and faint traces of carbon dioxide, helium, and neon. But once atmospheric oxygen falls below 19.5 percent, human cells begin to stop functioning. Below 14 percent, the brain starts to falter.

"So, our cutoff, our minimum threshold, is 14.5," Staats said.

"Obviously, if anybody feels short of breath, has headaches, heart rate increases . . . ," Adams said.

"What if I feel amazing?" Staats asked.

"It's not all about you, Kai," Adams said. Everyone laughed.

They agreed to turn the carbon scrubber on once carbon dioxide levels reached 3,000 parts per million (seven times the normal atmospheric level), although it was unclear how much oxygen would be left at that point.

A few moments later, Staats announced that it was time to begin. "Cute little things aside, it's time for science," he said. One by one, we stepped into the module, leaving a small crowd of onlookers outside. He checked to make sure everyone had water. "If you have to go to the bathroom, go now!" he said. Tresch had a bag of Doritos, and there was some disagreement over whether they could be eaten inside, because Staats couldn't stand the smell. Finally, Staats pushed the heavy white hatch closed and tightened the seal with a long metal lever. It was muggy and the air was still. We sat down around a plastic table, and waited.

The first life-support systems were created by NASA in the late fifties for Project Mercury, a series of suborbital and low Earth orbit flights that lasted up to thirty-five hours each. Mercury's capsules reached an altitude of nearly 200 miles, some seventeen times higher than the 60,000-foot point at which pressure becomes less than one pound per square inch and water boils at 98 degrees Fahrenheit, the temperature of the human body. Without pressurization, bodily fluids start to bubble and evaporate. To keep the Mercury astronauts alive, their capsules were equipped with eight pounds of pure oxygen and a system that metered it out to maintain a pressure of around 5 psi (atmospheric pressure at sea level is around 15 psi). The carbon dioxide they exhaled—over two pounds per astronaut per day—was funneled through canisters, where it reacted with lithium hydroxide pellets to become lithium carbonate; the canisters were later discarded. If they got hungry or thirsty, the astronauts could eat tubes of pureed beef or drink Tang.

Even in those early days of spaceflight, scientists understood that long-term missions would require the creation of self-sufficient bioregenerative systems. In the sixties, the USSR developed a 1,300-square-foot test chamber that connected by air duct to a quag of green algae, which consumed carbon dioxide and expelled oxygen. Researchers living in the chamber could eat the algae, but they found it so repugnant that Soviet scientists had to introduce other, more appetizing plants, such as wheat and cucumbers. The program grew to include a facility called Bios-3, an underground structure where cosmonauts lived for months at a time off the oxygen and calories supplied by crops grown under lamps; black-and-white photographs of the facility show tightly planted stalks of wheat, bent over like windswept beach grass. When scientists attempted to grow wheat on the Russian space station Mir in 1996, the lamps broke and the crop failed. The second crop seemed to thrive at first, but when researchers examined the specimens after they returned to Earth, the plants turned out to be sterile.

Attempts by the European Space Agency (ESA) to build a biological life-support system have similarly revolved around algae—namely *Arthrospira*, a cyanobacterium that looks like fusilli bucati pasta. *Arthrospira* is an efficient consumer of carbon: one hectare of it can fix over six tons of carbon, whereas a hectare of trees can fix four at best. The algae is also "very digestive," according to researchers, but given astronauts' distaste for it, the ESA loop features a food compartment where crops such as beets and lettuce are cultivated. Excrement in the system is liquefied anaerobically by high-temperature bacteria and gradually broken down into ammonium, volatile fatty acids, and, eventually, nitrate, which is used to fertilize the plants. This approach, highly compartmentalized with each chemical transformation contained in a dedicated, separate chamber, allows scientists to isolate the forms of life whose functions are useful to them. But after three decades, researchers have yet to enclose a human inside.

NASA has also dabbled in space agriculture. In the late nineties, it conducted experiments at the Johnson Space Center in Houston called the Early Human Testing Initiative, enclosing volunteers in sealed chambers for up to three months at a time. In one experiment, the oxygen for a single crew member was supplied by twenty-two thousand wheat plants. A more ambitious project to enclose four people, named BIO-plex, was planned for the early aughts, but was ultimately shelved because of budget concerns. Still, NASA researchers have continued work on space agriculture, albeit on a more modest scale. A few years ago, astronauts succeeded in growing lettuce aboard the International Space Station in a miniature garden called Veggie.

Most recently, the China National Space Administration has collaborated with Beihang University to build Yuegong-1, or Lunar Palace 1, a sealed structure with small apartments and two growing chambers for plants. Beginning in 2018, eight student volunteers lived in the capsule, rotating in groups of four for over a year. Their diet consisted of crops they grew, including strawberries, along with packets of mealworms fed with biological waste. Like the ESA's loop, carbon dioxide was cycled through plants, which were enriched with nitrogen from processed urine. Yet even Lunar Palace 1 fell short of being a truly closed system. While it managed to recycle 100 percent of its water and oxygen, it managed to do so for only 80 percent of its food supply.

As space agencies have haltingly investigated biological life-support systems over the decades, they have been joined by an increasing number of enthusiasts operating outside the purview of government research. The largest of these projects, and the only one that set out to establish a completely closed system, was Biosphere 2. Behind the project was a commune-living theater troupe accused of being a cult and bankrolled by the scion of an oil tycoon. Their leader, a charismatic beatnik named John Allen, had an MBA from Harvard and kept company with Abbie Hoffman. They were inspired by the work of the American architect Buckminster Fuller, whose theory of synergetics led them to depart from traditional scientific methods that aimed to isolate and understand single variables. Instead, they wanted to initiate multiple processes at once to see what might emerge.

Things began to spiral out of control almost immediately. Once the system was sealed, oxygen levels inside started declining and carbon dioxide levels began to climb. The volunteers tried addressing the problem by cutting down inedible grasses and preventing them from decomposing, and by filling every possible corner of the habitat with new plant life, but to no avail. With the supply of breathable air dwindling, they decided to inject liquid oxygen directly into their atmosphere.

Shortly after the first mission ended, scientists at Columbia University figured out what had happened. Among the thirty thousand tons of dirt brought into the enclosure were rich soils meant for the rainforest biome. These soils were teeming with microbes that had similar metabolic processes to humans. If anything, given all that additional animal life, carbon dioxide levels should have increased even more than they did, but calcium hydroxide contained in the enormous concrete foundation reacted to and captured carbon dioxide through a process called carbonatation. In so elaborate a system, it's nearly impossible to account for everything.

Improvisational, frenetic, and fueled by romantic ambitions of otherworldly life, Staats's project embodies in many ways the ethos of Biosphere 2. Throughout his renovations of the test module, Staats has spoken with members of the original crew, including Taber MacCallum and Jane Poynter, founders of a company called Space Perspective that aims to take tourists on high-altitude balloon rides to the stratosphere. The couple, who have long dreamed of traveling to Mars, is also slated to be the first to test a long-term enclosure of SAM once it is fully up and running. Staats joked that he would have chocolates waiting for them on their pillows.

After sealing the door, we all settled around a folding table. Staats and Adams started talking about Linda Leigh, who spent three weeks sealed inside the test module by herself as part of the original Biosphere 2 team. "I would rather fly to Mars and be seven months alone in a spacecraft than be with someone else," Staats said. It struck me as revealing that someone so interested in space exploration should also be so preoccupied with being alone. The stuff from which a closed ecosystem is constructed—the plants that breathe with us, the soils in which we decay, the bacteria and fungi and viruses that live inside and around us—underscores the interdependence that sustains life. So intricate are these relationships that they may never be perfectly rendered in simulation, even in simplified form. I thought about a conversation I had months before with Francesc Gòdia Casablancas, a chemical engineer who runs the ESA's pilot plant in Barcelona, one of the most complex biological life-support systems ever developed. He told me that no matter what, his systems would always lose efficiency over time; the simplified biological cycle built by scientists in a series of reactors would never be "a perfect world." Living on a planet that still harbors secrets seemed to me like the opposite of being alone. There is a strange kind of companionship in the tension of not knowing, in the fact that the systems supporting life on Earth operate beyond our control.

Over the next few hours, we tried to play a rocket launch simulation board game called Xtronaut, whose instructions and game play seemed indecipherable. After about forty minutes, Tresch suggested we do an Instagram Live video, for PR's sake, but Staats didn't have the app on his phone. He texted his girlfriend, Colleen, who helps manage the project's social media, to see if she could upload something. We checked the carbon dioxide levels. They were at 953 ppm, only a couple hundred higher than when we started. It was taking longer to increase than Staats expected, and he wondered aloud about how to speed things up. Tresch suggested that we turn the scrubber on at 1,000 instead of 3,000 ppm. Morgan said she was hungry. "We wouldn't be very good on a long-term space mission," Tresch said.

Outside, the afternoon sunlight began to soften. A documentary filmmaker friend of Staats's who was responsible for monitoring the lung reported that it had sunk by a couple of feet since we had started the experiment; the temperature inside had fallen by ten degrees—to 91—but outside it had fallen only by four. Around the two-hour mark, we measured the oxygen concentrations in our bodies. We were fine.

After more than three hours, carbon dioxide levels were slightly higher than 1,500 ppm. Tresch turned on the fans that would pull air through the scrubber's zeolites at 7,500 cubic feet per minute. Inside the carbon scrubber's metal encasement, the fans emitted a low hum. Minutes passed, and the sensors showed carbon dioxide levels continuing to rise. Staats began to consider questions they hadn't answered beforehand. At what density do zeolites function best? Are cylinders the best shape for the job? Was the carbon dioxide moving through the cylinders too quickly? They tinkered with the fans, turning up the speed to see whether it would accelerate the carbon dioxide drawdown. "What's the CO_2 right now?" Adams asked.

"1,561," said Morgan.

"It's not going down," Staats said.

"It's not going down," Morgan said.

We waited another hour, nervous and tired. The board game had been long abandoned. We started putting together a robot designed to farm crops, but left it only partly assembled; there weren't any beds that needed tending anyway. Nobody knew why the carbon dioxide levels had risen so slowly, or why oxygen stayed constant around 17 percent, or why the scrubber didn't work. We forget how much we don't know, Staats told me later; it's not bad to be reminded.

Outside, some graduate students had gathered in the arc of light around the test module to watch us through the glass. I thought again about the eight original volunteers, stuck for two years inside a world of their own making. They subscribed to the Gaia hypothesis, which posits that the planet is a single self-regulating organism; that the earth's conditions aren't just uniquely sweet to human life, but inseparable from it. They described a sense of harmony with the atoms that flowed through them and the surrounding plants. Once we entered the test module, though, that kind of harmony seemed like a distant dream. Instead, our chemistry was doing things we hadn't anticipated as we played at launching rockets. I wondered whether Staats was right, whether we as a species would be compelled to seek some new terrain, a foreign place where our appetites might grow, expanding and expanding, like the universe itself. It was after eight o'clock; we had been locked inside for four hours. Soon we would leave the new world, trailing our chemical effusions behind us, and we would go back into the old one, where stars freckled the dark atmosphere and the students waited with lemonade.

MARION RENAULT

A French Village's Radical Vision of a Good Life with Alzheimer's

FROM *The New Yorker*

FOUR YEARS AGO, I spent a morning cooking couscous with my grandmother Denise near Grenoble, France, where she has lived most of her life. We peeled carrots and turnips, seared lamb and chicken, tied bouquets of herbs, and mixed hot water into the grains with our bare hands. I wrote down her recipe as we went along. My *mamie* has Alzheimer's, and I had to learn to make her couscous on my own, before she forgot how to do it herself. That day, I recorded a video of her on my phone. She was sitting in a familiar kind of wooden Ikea chair that you have probably sat on before and that I will always associate with her. As she gazed out the window, a thought occurred to her, and she turned to me and asked, "*C'est samedi que tu pars?*" You're leaving on Saturday?

Yes, I told her. I was returning to the United States, where my parents moved our family when I was eighteen months old. I found it painful to leave; each time we visited France, the progression of her disease seemed to become more unignorable. Her pencil trembled when she practiced her handwriting. She moved her daily baguette from the kitchen counter into the plate drawer. Late at night, she muttered and puttered around her apartment. When her wandering inconvenienced us, we guided her back to her chair. My family talked about the chair as if it were her refuge; it was probably more accurately described as our refuge from her confusion.

In the summer of 2020, my grandmother stopped eating and getting out of bed. She had fallen, fractured a vertebra, and forgotten about it. I flew to France with a dozen of Mamie's favorite sesame-seed bagels, and I lived with her as she recovered, fetching prescriptions for the pain she was constantly rediscovering, and rubbing her back when she coughed until she retched. I lay in bed with her until she fell asleep. I fed her. I learned, for the first time in my life, what it meant to care for someone. After five weeks, my mom took my place as Mamie's at-home caretaker.

Like so many families that are affected by Alzheimer's, we searched desperately for a new place where my grandmother could live. We viewed her isolation in her seventh-floor apartment as a risk to her health and safety, and felt that it was not only right but necessary to exchange what was left of her autonomy for the round-the-clock, structured care that she could receive at a nursing home. About half of the six hundred thousand people who live in France's EHPADS, or "housing establishments for dependent elderly people," have dementia. These are imperfect institutions: in 2018, French nursing-home workers went on strike to protest staff shortages and cost-cutting, and, earlier this year, disturbing reports of abuse and neglect, untrained staff, and the rationing of food and diapers by a for-profit nursing-home company put the country's elder-care system under intense public scrutiny.

A nursing home in a nearby suburb finally offered her a place after weeks of uncertainty.

My grandmother's life now seems safer but smaller. Her memory-care unit is locked with keypads to prevent her from wandering out and is rarely unsupervised; the woman who taught me to cook couscous no longer has a kitchen. My family is satisfied with her care: the staff is affectionate, Mamie is often cheerful during visiting hours, and she regularly participates in Montessori activities such as vegetable peeling and sing-alongs. I never saw her with pets when I was younger, but she now lets the nursing home's service dog, an enormous Labrador named Nova, cuddle with her in bed. Still, it seems inevitable that, as my grandmother's condition declines, she will lose the few freedoms she has left. Last year, I stopped bringing her bagels after I noticed that they were furring themselves green inside their plastic bags. This year, she complained of being weaker, of fighting with her brain but not understanding why. She sometimes referred to her nursing home as her aunt's house, or the children's daycare where she worked for decades. Someday soon she will no longer be able to play dominoes with me—she won't understand how to win, or even how to count the dots on each tile. Later, she might be moved to the unit next door, where people with more serious cognitive limitations live under even closer surveillance.

Anyone who has cared for someone with Alzheimer's is likely to be familiar with this transaction. We cede their freedom to gain a sense of security—theirs, but also ours. We attempt to resize their world, removing the choices that might pose a danger to them. But I often wonder whether the standard approach of a nursing home—the constant surveillance, the rigid schedules for waking, bathing, eating, socializing, and sleeping—is the best that we can offer to loved ones with dementia.

This summer, before going to see my grandmother in Grenoble, I visited a nursing home that aims to expand, not restrict, the liberties of people with Alzheimer's. The Village Landais, situated in Dax, in southwestern France, is part of a movement to make memory-care units less like hospitals and more like small neighborhoods. Some of these facilities are designed to convince residents that little has changed—"That life is still as it was once, with children to take care of, and holidays at the seashore, and familiar homes to return to," as Larissa MacFarquhar wrote for *The New Yorker* in 2018. But the Village seemed to convey a slightly different message: that life remains full of choices and that autonomy enriches life. Its residents can come and go from their homes as they please, whether through the unlocked door or through a window. They can wake and shower at their leisure; they can shout, pilfer sweets, make tea at 2 a.m., sweep with the broom upside down, and handle sharp knives in the kitchen. Advocates for this kind of care argue that, for people with Alzheimer's, the risks of institutional dehumanization are just as profound as the physical dangers of cutting one's hand, or falling and breaking a bone. "Their cognitive troubles don't permit them to adapt to our world," Gaëlle Marie-Bailleul, the Village's head of medicine and a specialist in neurodegenerative disorders, told me. "We adapt to them." Most nursing homes devote themselves to the narrow and perfectly reasonable goal of keeping residents safe and healthy. The Village

Landais contemplates a broader question: What might a good life with Alzheimer's look like?

A hundred and eight people, of whom the youngest are fortysomethings and the oldest are centenarians, live full-time on the seventeen acres of the Village Landais. Its sixteen group houses are clustered into small neighborhoods, and each house features two staff members, who are trained in disciplines such as home care, occupational therapy, and gerontology. Living areas are filled with natural light and secondhand furniture. Hallways are designed as loops, without dead ends, to reduce confusion; each resident has her own bathroom, with a mirror that can be folded up when she no longer recognizes her own reflection.

In the Village's restaurant, which is supposed to open to the public in 2023, I met Nadine Zoyo, who began baking as a child during the war and spent many years as a homemaker, catering the baptisms and weddings of loved ones. "She never stopped," her daughter Béatrice, who was dining with her, told me. As her Alzheimer's progressed, Nadine struggled with words, and repeatedly fell and injured herself. Because Béatrice could see her mother's apartment from her own, she told me, "I was always looking out the window." The Village seemed to reanimate Nadine: she used to sit still for long periods, Béatrice said, but now she knits, interacts with others, and seems to lose her train of thought less often. "It's extraordinary," Béatrice continued. "She is living again."

In many respects, the Village lives up to its name. Residents can tend to a large garden each morning and feed Junon and Jasmine, two donkeys who keep the grass in check. A salon offers haircuts, and each house makes a daily grocery run to the *épicerie*, or supermarket. The store has no cash register or price labels, however; the cognitive work of budgeting and paying has been conveniently edited out. The Village's hundred and twenty employees, along with sixty active volunteers, travel the grounds on foot, or on bikes that are parked haphazardly around the campus. The medical staff do not wear lab coats; they conduct house calls, not office examinations. Marie-Bailleul told me that she tries to set aside expectations when she speaks with villagers during meals and walks. "They oblige us to be sincere, spontaneous, and in the present moment," she said. Does it matter if a patient mistakes her for a friend or a grandchild? If someone wants to eat yogurt with a fork, so what? As long as they have an appetite and feel cared for, these are positive experiences, she said.

One of the most radical aspects of the Village is its insistence that a person with Alzheimer's is not just diminishing into the sum of her symptoms, but flourishing and evolving as a human being until the end. Leticia, a forty-one-year-old villager with early-onset Alzheimer's, is learning to play the guitar. Many residents who never previously engaged in the arts take to painting or collage-making, staffers told me, and former marathoners and cyclists can re-create long runs and rides within the village. (Academic researchers have noted that some people with dementia appear to enjoy enhanced artistic abilities; Mary Mittelman, a research professor at New York University, told me that, in the chorus she founded for people living with dementia and their families, those who may not remember what they ate for lunch are able to learn as many as eighteen new songs for each concert.)

A bright green train car sits in the Village library, hitched to nothing in particular. A therapeutic tool, its interior is realistic, with metal racks for baggage and a flat-screen television, which plays footage shot from a train as it rolls through a forest. Nathalie Bonnet, a staff psychologist, told me that the simulacrum of travel appears to quell a simple desire to be elsewhere: she has seen agitated villagers fall asleep on the car's cushy seats, or sit and articulate worries that they could not before.

Bonnet, who has silver hair and was wearing earrings shaped like droplets of water, led me to a terrace in one of the Village's little neighborhoods and explained its philosophy. "As long as they *can* do, we must be able to leave them the liberty *to* do," she said. "The spirit of security—of safety as a means to live longer—should be reconsidered. It's not about opening up all freedoms, either. It's not that. It's, 'What is the tolerable level of freedom to let the person live?'" Villagers can set the rhythm of their own existence, hour by hour, minute by minute. They can wash their own clothes, gather beneath expansive eaves, and walk unsupervised along looping wooded paths.

As Bonnet and I talked, a pair of residents ambled by. She asked a gloomy-looking woman, "How are you, Claudine?" Claudine, a former hairdresser, shrugged morosely, tugging at her sweater and pant pockets.

Bonnet asked again how Claudine was doing. Sensing that something was wrong, she rose out of her seat and took the woman's hand.

"You're looking for something?" Bonnet asked softly.

"Excuse me," Claudine said sorrowfully, unable to explain.

"It's all right," Bonnet said, her voice softening even more. "I've got plenty of time."

Time and intimacy are especially precious in understaffed nursing homes, and in families that care for those with dementia. Despite myself, I'd often felt irritated when I had to stop the clock in my world in order to accompany my grandma in hers. In Bonnet, I saw no sign of irritation. She asked Claudine whether she was worried that someone had taken her belongings. Claudine nodded, so Bonnet, still stroking Claudine's hand, suggested that she go check that her bag and coat were safe in her room.

"There's the style of communication where you have few words," Bonnet told me as Claudine walked off. "We find a way of decoding."

The Village's operating costs exceed six million euros a year, of which about two-thirds come from public coffers. In exchange, researchers are studying the experiences of Villagers, from their behavioral troubles to their medication use and levels of depression and anxiety. "It does not suffice to want to do well," Hélène Amieva, a researcher and professor of gerontology at the Université de Bordeaux who is independently studying the Village, said. The Village seeks to demonstrate that its philosophy of elder care has measurable positive impacts—that the day-to-day quality of life of its residents improves, or that their disease progresses more slowly. Research into medical outcomes is still ongoing, although a survey has suggested that, since the Village opened, members of the public who live nearby have formed more positive associations with Alzheimer's, and may see those with the disease as warmer and more competent than they previously did. Another group of researchers is studying economic feasibility. Some families with financial need pay as little as three thousand euros a year, but others pay up to twenty-four thousand—and even that is not enough to cover the majority of the Village's operating costs. It remains to be seen whether medical savings—for example, in the form of fewer hospital visits or reduced medication use—will offset some of these expenses.

In the US, where one in four nursing homes faces employee shortages, experts were skeptical that such a model could ever be implemented on a large scale. "That kind of staffing is not even there in our ICUs," Joe Verghese, a neurologist and the chief of geriatrics at Montefiore Health System, told me. Elena Portacolone, an associate professor of sociology at the University of California, San Francisco, went so far as to reject the Village's basic design, and argued instead that Alzheimer's patients should be integrated into society. "To me, it's segregation," she said. "I think it's wrong." Manon Labarchède, an architect and sociologist who recently completed her PhD dissertation about Alzheimer's, at the Université de Bordeaux, said that, if the village model remains closed off from the outside world, it will fail to change societal views of the elderly. Still, she said, it helpfully explores an alternative to traditional nursing homes. "It shows other things are possible."

Dementia isn't unique to our species—it also shows up in dogs, cats, horses, and rabbits—and has probably been with us for centuries. In a cultural and medical history of dementia, *Dementia Reimagined*, the psychiatrist and bioethicist Tia Powell notes that the writer Jonathan Swift is thought to have been afflicted by it in his old age, during the eighteenth century, when he complained of a fleeting memory, an ill temper, and a lasting despondency. "I have been many months the shadow of the shadow of the shadow," he confessed in one letter. In another, he told his cousin, "I hardly understand a word I write." When Swift died at seventy-seven, in 1745, dementia was seen less as a medical condition than as an inevitable feature of aging or, in some cases, a kind of madness. Not until 1906 did Alois Alzheimer, a German pathologist, argue that one of his patients had lost his memory because of a tangle of proteins identified in his postmortem brain. Over the years, studies have suggested that Alzheimer's causes at least 60 percent of dementia cases.

Americans long dealt with dementia by institutionalizing the people who experienced it. In the time of Alzheimer, the US housed them in cramped poorhouses, where they frequently came down with infectious diseases, and suffered chronic neglect and abuse. One 1909 report describes a Virginia poorhouse warden who stopped an older woman from wandering by anchoring her with a twenty-eight-pound ball and chain. Eventually, poorhouses were replaced by mental hospitals, and mental hospitals were replaced by nursing homes. These facilities were a step forward, but they limit autonomy by design, and they often overuse antipsychotics as chemical restraints.

Dementia finally came to be seen as a public health crisis in the late 1970s. In 1976, the National Institutes of Health spent $3.8 million on Alzheimer's research; by the year 2000, federal funding for research on Alzheimer's and other types of dementia had reached $400 million. But this money has overwhelmingly been spent on trying to eradicate Alzheimer's, and not on experiments in dementia care, like the Village. Even the Alzheimer's Association, the country's leading advocacy group for people with the disease, envisions "a world without Alzheimer's," rather than a world in which we try to live with it peaceably. But the dream of vanquishing Alzheimer's has proved elusive. Alzheimer's drug trials almost always fail. In June 2021, the US Food and Drug Administration approved aducanumab, the first novel Alzheimer's drug in almost twenty years—against the recommendation of an advisory panel, which overwhelmingly concluded that there was insufficient evidence to deem the drug effective.

"Because the drugs keep failing, people are, like, 'What do we offer people?'" Kristine Yaffe, a neurologist and psychiatrist at the University of California, San Francisco, told me. "What do we say to our patients?" In the US alone, some 6.5 million people over sixty-five, of whom a disproportionate number are women and people of color, already have Alzheimer's. Between one-third and one-half of Americans aged eighty-five or older are estimated to have dementia. Most people live between three to eleven years after an Alzheimer's diagnosis; some survive for decades. Because of a shortage of elder-care infrastructure and workers, many of them will face the disease with far too little support. "We're not prepared," Esther Friedman, a University of Michigan sociologist who studies elder care, told me.

News coverage of dementia is far more likely to focus on how to prevent it, or how much it burdens our health system, than to highlight the experiences of people who live with it. In surveys, many adults report fears that, if they were diagnosed, they might lose their health insurance, driver's license, or job. More than half expect a person with Alzheimer's to lose the freedom to make their own medical decisions, as my grandmother eventually did. "It's a disease that scares, and that repulses," Marie-Bailleul told me. As the sociologist Karen Lyman has written, people with dementia are often depersonalized into "merely disease entities." Powell notes in her book that, in 2007, a bioethicist even explored the philosophical argument that because dementia destroys personhood, a person who develops dementia has a moral obligation to kill herself. "Not killing herself would show selfish callousness," he wrote. "She causes unnecessary harm to others by imposing significant burdens on them rather than autonomously solving the problem."

Our fear and hatred of Alzheimer's ultimately seems rooted in our modern attachment to the idea of the self. "The self is also a creation, the principal work of your life, the crafting of which makes everyone an artist," Rebecca Solnit writes in *The Faraway Nearby*, a memoir that touches on her mother's Alzheimer's, among other subjects. "She was herself being erased." By yoking our humanity to our cognition, however, we risk dehumanizing those whose grasp on memory, language, and perception slackens. Families may stop bringing loved ones with Alzheimer's to restaurants and gatherings; they may take away quotidian things, like the freedom to run errands or set the table or have neighbors. This summer, with a twinge of guilt, I realized that my family did not refer to Mamie's nursing home as her home, but rather as "*là, ou elle est*"—there, where she is. "We consider that, because you have lost your memory, you are incapable of anything," Pascale Lasserre-Sergent, the director of the Village Landais, told me. "We consider that you no longer exist as a person." The poet Tony Harrison wrote:

If we *are* what we remember what are they . . .
who, when evening falls, have no recall of day

In recent years, new philosophies of memory care have emerged. In 2009, a dementia village called the Hogeweyk opened in the Netherlands, funded mainly by the Dutch government. Its houses evoke various Dutch lifestyles—one is for urbanites, another is for culture lovers, and another is for people with religious affinities—and residents can visit a pub, restaurant, theater, and supermarket. The Village Landais is not affiliated with the Hogeweyk, but the Village's press representative, Mathilde Charon-Burnel, told me, "Their example inspired us." The Village, she said, aims to take the ideal of autonomy even further, and to scrutinize its own impact by using a scientific approach. Even if researchers discover that the Village fails to improve patient outcomes, Charon-Burnel told me, they will still publish their results so that other organizations can learn from them.

These alternative approaches do not pretend that the disease is anything but cruel. Alzheimer's takes away so much that we consider essentially human: knowing, remembering, expressing. But Bonnet, the psychologist, pointed out that people with Alzheimer's often show a gift for rich presence that eludes many of us. When patients forget about their own condition, a development called anosognosia, they sometimes feel better, as my grandmother did. They inhabit the present moment and may let go of troubling memories or fears about the future. Even as their experience of the world is transformed, they find ways to describe it that the rest of us can understand. "It can be very imaginative, very symbolic," Bonnet said. A resident might tell her that they took a flight to go grocery shopping; it felt like a long journey. If someone tells her "I saw my mother," she understands that someone took special care with them. Marie-Bailleul told me about a conversation she had with a woman who was grieving the loss of a fellow resident. The woman pointed to the leaves of a tree, which were riddled with holes. "Look, he's crying," the woman said. "He lost his friend."

My family tries to remember the things that my grandmother has forgotten. Her father fed seven children by farming someone else's land. When she was twenty, she married my grandfather Angelo, a French Italian man twelve years her senior, who had survived eighteen months in a labor camp during the Second World War. Angelo worked at a bottling plant and later became a welder; Denise worked at a factory that made metal bearings, then at a day care.

My grandfather developed Alzheimer's before my grandmother did. She knew the complicated feelings of pity and protectiveness, the uneasy impatience, the sweetness and sadness, of caring for a loved one with dementia. When they visited my parents in the US, he went out for a long walk while mistakenly wearing her size 5 shoes, and she waited by the front door, fuming. She got embarrassed when he mistook the curtains at Olive Garden for giant hanging napkins. She was his primary caretaker until he died at home in France, from a lung infection after aspirating food. Then, within a year or two, she began to experience symptoms of her own. "I'm sick of this disease," she cried out one morning from her chair, according to a journal I kept at the time. "What did I do to God that he did this to me?" More than once, she told me that Alzheimer's was devouring her life. I grieve not only for the life she is forgetting but also for the hardship it has contained.

My grandmother will probably never relocate to a place like the Village, but I have started to wonder whether I have the power to bring parts of the Village to her. How would my grandma choose to spend time with me if I allowed her to set our itinerary? Are there new hobbies or activities that I could invite her to explore—or old ones, like cooking, that I could reintroduce with a simple gesture, like bringing her potatoes or carrots that we could peel together? "Discovery is possible in this disease," Bonnet told me. Solnit, in *The Faraway Nearby*, wrote about how Alzheimer's drew her closer to her mother: "In that era, I think my voice and other things registered as familiar and set her at ease, and perhaps she knew me more truly. And perhaps I her, as so much that was superfluous was pared away and the central fact of her humanity and her vulnerability was laid bare."

During my latest trip to see my mamie, I asked her, "Who am I to you?" She paused for a moment, then smiled. "You are my little sister," she said. Once I might have fretted about how far from reality she had strayed. But this time I tried to share her interpretation of reality, instead of imposing mine onto hers. What does it mean to be a younger sister? It means that I am someone who has giggled, cried, cooked, and played with her. It means that she has protected me, and that she feels that she is safer when I am near. My grandmother's answer was not accurate, but it was truthful.

A few days later, on my last visit of the summer, I found myself unable to say goodbye. I searched my mind for words that would capture how easy and familiar it felt to be with her, and that I wished I were not going, and that, as soon as I could, I would be back. "See you tomorrow," I told her. "*À demain, ma chérie*," she replied. See you tomorrow, my dear.

In El Salvador and Beyond, an Unsolved Kidney Disease Mystery

FROM *Undark*

JOSÉ LÓPEZ DIDN'T want to die, but the alternative—having a scalpel plunged through his abdominal wall to install a soft, silicone dialysis catheter—filled him with terror. For weeks in the fall of 2021, the then-thirty-four-year-old agricultural worker from Tierra Blanca, El Salvador, had refused the surgery, holding out instead for a miracle from God. Regional lore held that such acts of grace were possible: There was the man from Las Salinas whose invocations had restored his ailing kidneys; the boy from La Noria who was recovering swiftly after devoting himself to the gospel. Through his mounting illness, López clung to the rumors and prayed for a similar deliverance.

But he was running out of time. The fluid buildup in his abdomen had grown so severe he felt like he was choking. He couldn't stand, eat, or sleep. His legs had gone completely numb, and he felt phantom ants crawling across his palms. He had lost nearly half his body weight, and sharp knobs of bone poked through his skin like tentpoles. The primary indicator of kidney function, creatinine levels, are considered high when they reach 2 milligrams per deciliter of blood. By 6 mg/dL, they indicate severe, life-threatening kidney disease. López's were at 35.

Doctors had delivered a stark ultimatum: Begin dialysis or die. But López had seen dialysis firsthand. He was the eighth man in his family to develop chronic kidney disease of unknown origin, or CKDu, a beguiling and fatal illness that was decimating entire communities across Central America and other warm agricultural locales around the globe. Just in the last three years the disease had claimed his beloved father, Vitelio, and two uncles. The only adult López men who remained were José and his two younger brothers, each of whom had the disease and were inching toward similar fates. Many of José's relatives had spent their last months receiving dialysis treatments before eventually succumbing—José understood that the catheter was the end of the line.

Nevertheless, his then-four-year-old son, José-Vitelio, finally broke through to him. The whole family—his mother, María Luisa; his wife, Marta; his then-thirteen-year-old son, Edwin—had gathered at his bedside and were begging him to reconsider when, López recalled, the child piped up: "We don't want to be left alone," he said. The tiny plea nudged López to seek help. The family called a car, and López was driven to the San Juan de Dios National Hospital in nearby San Miguel. The catheter was installed under local anesthetic, and several days later López enrolled in the Jiquilisco municipality at-home peritoneal dialysis program.

"Here I am, still fighting," he said, seven months after the procedure.

Since the 1990s, tens of thousands of people across Central America, Sri Lanka, India, and elsewhere have been killed by CKDu in a ballooning epidemic that has baffled researchers, overwhelmed health care systems, and wiped out entire families. Agricultural communities seem particularly vulnerable, but the disease has surfaced among other workers, too. The precise number of deaths is unknown, but in locales where the disease is endemic, such as the Bajo Lempa region of El Salvador, experts estimate that up to one quarter of the male population has CKDu. (It affects men at roughly two times the rate of women, according to the International Society of Nephrology.) It cannot be cured, only treated with dialysis, and in rural communities such as Tierra Blanca, it is rare for dialysis patients to survive more than a few years. But despite its devastating toll, scientists have struggled to determine what causes the disease or how it develops.

Two competing hypotheses have emerged to explain the epidemic. The first suggests that the main cause is an unknown toxic agent—pesticides, perhaps, or heavy metals or silica. The second points to heat stress and dehydration, amplified by brutal working conditions and an increasingly warmer climate. Many researchers have accepted the possibility that the disease may emerge from the synergistic effect of two or more causes working in tandem. Yet the debate over the primary driver of the epidemic—toxins or heat stress—remains unresolved.

The lingering scientific uncertainty holds profound implications for policy, prevention, and treatment. Without a clear scientific consensus, governments, nongovernmental organizations, and doctors are divided over how to best coordinate a response to the epidemic, a gridlock that has sometimes flared into outright hostility: At least two individuals involved in CKDu research and policy claim to have received death threats and violent attacks because of their positions. In recent years, however, studies from both sides of the debate have offered intriguing new clues, giving scientists renewed hope that someday the scientific stalemate may be broken, and the mystery of this deadly disease finally unlocked.

Until then, the only weapon CKDu patients like José López have against the disease is dialysis. Of the two varieties of the treatment—hemodialysis, which requires an expensive machine to remove toxins from the blood, and peritoneal dialysis, in which the patient's abdomen is flooded with a solution that leaches toxins from the abdomen's blood vessels—López can access only the latter. The treatments are relatively inexpensive and can be performed at home, but it must be done four times a day. His son Edwin administers the treatments. López wants Edwin to see the disease close-up, as he had with his own father. He hopes it will encourage the boy to study hard in school, so that he might avoid the brutal agricultural labor that López believes is responsible, in one way or another, for his illness.

"You have an example, right in front of your eyes," López often tells his son. "Look at what has become of me. This is how the fields left me."

Tierra Blanca, a small agricultural community of several thousand, is nestled within the verdant coastal lowlands of El Salvador's Bajo Lempa region. Nearby, a few squat volcanoes rise above the haze of burning sugarcane fields, and to the south the seafood-rich estuaries of the Bay of Jiquilisco give way to the open Pacific Ocean. West of town, the Lempa River braids listlessly through endless tracts of farmland, depositing the mineral-rich silt that makes this region one of the most fertile in the country.

López began cultivating the land around Tierra Blanca when he was twelve years old. His father, Vitelio, and grandfather, Juan-Francisco, would load the boy into an oxcart before sunrise and go clattering down the rocky trail that led into the fields, where roads had not yet been laid. They rented a small subsistence plot from some landowners in San Marcos, and there López learned how to farm—how to drive the oxen and mix an organic pesticide, measure seed depth with his thumb, and trench canals to drain the monsoon rain. Although López was a good student, school supplies were expensive, and he dropped out in the seventh grade.

Back then, in the late 1990s, kidney disease had not yet been widely identified as a killer of El Salvador's workingmen and -women. The country had just emerged from a brutal twelve-year civil war that had claimed the lives of seventy-five thousand people, mostly civilians. In the Bajo Lempa, the health care system—insofar as it existed at all—was privatized and costly. A common saying in the region at the time went, "You pay or you die," and most of the men López worked with in the fields simply could not afford to pay. When they died, very few people ever questioned why.

That began to change in the late 1990s, when Julio Miranda, a local community organizer and health care activist, started asking questions about the excessive number of funerals that were taking place in Tierra Blanca. "People were dying one after another," Miranda recalled. "And without any sort of diagnosis." But Jesús Domínguez, a Spanish doctor who had come to El Salvador to render medical care during the civil war and who worked closely with Miranda, started to recognize the shared symptoms of the deceased. "It was clear," he said, speaking over Zoom from France, where he currently lives. "They were dying from kidney failure."

In the early 2000s, Miranda, Domínguez, and a handful of other activists began collecting urine samples from field workers over the age of thirty, looking for the elevated protein levels that often signaled kidney disease. Individuals with worrisome levels were instructed to go to a lab for blood work. Dominguez still vividly recalls the night the first batch of test results came back from the lab. The team stayed up all night in Miranda's living room, poring over the results, checking them again and again. Finally, in the morning, Domínguez recalled turning to Miranda. "This is a massacre," he said.

"We had the death sentences of hundreds of people in our hands," Domínguez said in a recent interview, later adding: "I was in shock."

As the scope of the emerging epidemic began to dawn on the activists, Miranda immediately set about establishing a registry of everyone in the area who was showing signs of kidney disease. The data, he hoped, might galvanize the health ministry into investigating the situation. This registry—which included as much medical information as possible about each patient—was new.

But in a separate notebook, Miranda had already begun to list the dead.

One morning in June of 2005, José López entered the bedroom of his grandfather, Juan-Francisco, to rouse the older man for work. Although Juan-Francisco had seemed perfectly healthy just the day before, López now found him gravely ill. He was too weak to stand, and after several failed attempts to get out of bed, he simply hung his head and began to cry. "The man upstairs is calling me," López recalled his grandfather saying.

Juan-Francisco died three weeks later. The family paid for a modest funeral service that included chicken tamales and black coffee, and a hearse to lead the funeral procession to the Tierra Blanca cemetery. While mourners sang religious hymns, the body of Juan-Francisco was interred in an unadorned cement tomb with ample space surrounding it—the family plot was sparsely populated then. He was the first López man to die of CKDu, and the first from their family to be added to Julio Miranda's list of the dead: Number 62.

The campus of Rosales National Hospital, the centerpiece of El Salvador's public health care system, sprawls across several city blocks in the heart of the nation's capital, San Salvador. Throngs of patients, staff, and students jostle between the tall, hangar-like buildings, while shotgun-toting security guards stand at each of the hospital's entrances. Rosales was built in the late nineteenth century with money from a public lottery, and a spirit of chance still lingers: Patients complain of erratic care and long waits, doctors of chronic underfunding and insufficient personnel. And nowhere on the campus is more teeming with fringed nerves than the kidney ward, which according to the head of nephrology, accounts for some 60 percent of the hospital's admissions.

If Rosales is chaotic now, it was far worse in 1995, when Ramón García-Trabanino arrived as a seventh-year medical student, ready to begin his hospital internship. Back then, there was no dedicated nephrology ward, only an internal medicine wing that housed a few old hemodialysis machines. When he started at Rosales, the young doctor supposed he would be treating the full range of conditions he had studied in school—strokes, neurological disorders, pneumonia. Instead, the unit was flooded with kidney patients, wave after wave of them, many on the brink of death.

"The amount of patients was just overwhelming," recalled García-Trabanino, who now, at fifty-one, directs a private dialysis clinic in San Salvador. "It was like an erupting volcano."

In medical school, García-Trabanino had learned the textbook profile of a kidney disease patient: old, diabetic, hypertensive, equally male and female. He quickly noticed that the influx of sick people arriving at Rosales did not conform to that profile at all.

"They weren't diabetic. They didn't have high blood pressure," said García-Trabanino. "They were in their twenties, thirties, forties—they were young. They attended once, twice, and then they died."

The patients were treated with an antiquated technique called rigid-catheter peritoneal dialysis. A stiff tube was inserted into their abdomens and their peritoneal cavities flooded with a toxin-absorbing fluid, a process that, to be effective, should have been performed daily. Because of the overwhelming demand, patients at Rosales only received one treatment per week, and each time they returned a new catheter would have to be installed. This alone was dangerous—if doctors accidentally nicked the patient's bowels or liver during the process they could die, and frequently they did. García-Trabanino, who says he still has nightmares from this time, vividly remembers the first patient he lost. His name was Ramón, just like him.

Perhaps most frustrating of all for the young doctor was that nobody could explain what was happening. Even Ricardo Leiva, the chief nephrologist, was dumbfounded. "We kept asking ourselves, 'Where are they coming from?'" recalled Leiva. "It was overwhelming the hospital. So that's why we did the first research."

In the fall of 1999, García-Trabanino, Leiva, and a few others began to investigate. They conducted interviews with 202 patients, collecting demographic, occupational, geographical, and clinical data. They found that only one-third of the interviewees had known risk factors for kidney disease such as diabetes or hypertension. The rest had "unusual characteristics that were not associated with the known risk factors," according to the resulting study. Of this latter group the majority were men, worked in agriculture, and had been exposed to pesticides. They also came from hot coastal areas like the Bajo Lempa region, where Tierra Blanca is located.

Published in 2002, the study concluded that a novel group of end-stage kidney patients had been identified, who seemed to "lack a cause for their disease." The researchers suspected a relationship with occupational exposure to agrochemicals, but stated that further research was needed. It was the first description of the emerging epidemic in the scientific literature, and for his efforts García-Trabanino was awarded the National Medical Investigation Prize, the country's top honor for medical research.

In the years preceding the disease's identification, pesticide use had been on the rise in Central America. According to the Pan American Health Organization, or PAHO, which serves as the World Health Organization affiliate for the Americas, by the end of the twentieth century, the region had the highest per-capita use of pesticides of anywhere in the world. Between 1994 and 2000, imports of pesticides to Central America rose by 32 percent. The average agricultural worker in Central America in 1992 used 9.9 pounds of pesticides per year. By 2000, that number had risen to 14.8 pounds. The most commonly sold pesticide in El Salvador, paraquat, has been banned by dozens of countries because of its acute toxicity to the lungs, liver, and kidneys. And although a link has not been conclusively established between paraquat and CKDu, it remains one of the prime suspects the toxic-exposure researchers are investigating.

Nevertheless, the pesticide hypothesis encountered an early stumbling block. In 2005, García-Trabanino's team published a study that compared CKDu prevalence in a population of men (mostly agricultural workers) in the Bajo Lempa, which is near sea level, with men in Sesori, which is farther north and higher in altitude. The findings perplexed the researchers. Although the two populations cultivated the same crops, using the same pesticides, the men at sea level were eight times more likely to have CKDu. It was a serious blow to their initial hypothesis. "It was not related, at least not to pesticides," said García-Trabanino. "It was related to the region. So we had to come up with new ideas."

Meanwhile, other hot spots were gaining attention. In neighboring Nicaragua, Catharina Wesseling, an epidemiologist who then directed an occupational health and safety initiative called the Program on Work and Health in Central America, began hearing reports of a fatal kidney disease impacting workers at a sugar mill near the town of Chichigalpa. In Sri Lanka, doctors in Anuradhapura District, in the North Central Province, were beginning to notice an alarming uptick in kidney patients arriving at their hospitals. In the state of Andhra Pradesh in India, and in the El-Minia Governorate in Egypt, similarly worrisome trends were coming into focus. Locales where the disease appeared endemic generally shared two conspicuous features: They were heavily agricultural, and they were hot.

The government in El Salvador, despite having awarded García-Trabanino the National Medical Investigation Prize in 2000, had largely ignored the epidemic in the decade following its emergence. That began to change, however, when in 2009 the country elected its first leftist president, the Farabundo Martí Liberation Front candidate, Mauricio Funes, putting an end to decades of conservative rule. Among the new president's priorities was a reboot of El Salvador's public health care system, which had atrophied under years of neoliberal policies designed to encourage privatization. To oversee the reforms, Funes appointed a new health minister, María Isabel Rodríguez, a doctor who had worked for PAHO and was keenly aware of the unfolding health crisis. Under Rodríguez's leadership, a young nephrologist named Carlos Orantes began to spearhead the first official government investigation into the illness.

In the fall of 2009, Orantes and his team conducted the largest CKDu study undertaken up to that date, surveying 775 individuals in the Bajo Lempa region in order to better understand the disease's prevalence and associated risk factors. Many of Orante's findings—such as the conspicuous absence of traditional risk factors, the higher burden among men than women, and the preponderance of the disease in agricultural communities—dovetailed closely with those of García-Trabanino and others from previous years. Unlike García-Trabanino and Wesseling, however, who by then had begun exploring other possible causes, Orantes determined that toxins, especially pesticides, deserved renewed scrutiny. Of the men that responded to his survey, 82.5 percent reported contact with agrochemicals, the highest percentage of any potential risk factor screened for by the study. The report noted that field researchers involved in gathering data had observed dangerous agrochemical handling practices and had become aware of a long history of indiscriminate aerial crop dusting in the Bajo Lempa region. The study concluded that certain risk factors "may act synergistically," and that although the specific etiology remained elusive, the hypothesis of an environment toxin "cannot be ruled out."

The study, with its renewed suspicions of a toxin-based etiology, formed the basis of the new health minister's stance on CKDu. "The evidence suggests agrochemicals and pesticides as possible associated factors and this has to be investigated," said Rodríguez in a 2013 interview with *MEDICC Review*, a Latin American health journal. Press coverage of the issue around this time tended to follow the government's lead, suggesting in article after article that there was a potential link between pesticides and the epidemic. During this time, researchers like García-Trabanino and Wesseling say they began to worry the public was being served a convenient narrative, one that provided an easy answer—but wasn't supported by the science.

In April of 2013, El Salvador hosted a meeting of the Council of Ministers of Health of Central America and the Dominican Republic, which culminated in the signing of the Declaration of San Salvador, which recognized CKDu as "a major public health problem," and signaled a willingness to tackle the epidemic on a regional level. Before the document was signed, however, a vigorous debate broke out amongst the attendees over whether the role of pesticides was being overstated. According to an account of the meeting by the Center for Public Integrity, representatives of El Salvador's ministry of health asserted that the most compelling data suggested a high association between CKDu and agrochemical exposure, while other researchers, including García-Trabanino, argued that a definitive link had never been established in the scientific literature. Rodríguez put an end to the debate by declaring, "What has been presented here is scientific fact, and I will defend it with my nails!" She showed the crowd a set of brightly painted red nails, the room erupted with laughter, and the document was signed.

But the debate over the role of pesticides did not end there. In the summer of 2013, Rodríguez and Salvador Menéndez, the mayor of a municipality called San Luis Talpa that had been devastated by the epidemic, launched a campaign to prohibit the use of fifty-three agrochemicals that were suspected of being deleterious to workers' health. The proposal ignited a fierce public discourse, whose fault lines hewed closely to what had been argued by each side in the Council of Ministers meeting. But as the scientific debate entered the public realm—through press coverage and word of mouth—the conversation quickly soured, before finally exploding into outright acrimony and accusation.

The mudslinging went in all directions: García-Trabanino accused the health ministry of distorting science in order to score a tidy political victory. The powerful agricultural sector, according to a newspaper article at the time, made sensational claims that El Salvador's agricultural output would plummet by 80 percent if the ban went through. Orantes says he was maligned as an "eco-fanatic" who was "obsessed with agrochemicals." In turn, García-Trabanino says he was publicly disparaged as being a shill for the agricultural sector (he insists he never took money from agricultural groups) and was so widely vilified that Julio Miranda had to warn him against returning to Tierra Blanca, where he had been conducting research. The field workers whose health had been the focus of his career were now a threat to his safety. García-Trabanino says he received a death threat during this time. For Mayor Menéndez, it was even worse: He claims that on three separate occasions, the armored car he was riding in was fired upon by unknown assailants. Each time, he notes, he escaped without injury.

On September 5, 2013, El Salvador's legislative assembly voted to approve the agrochemical ban. But in a move that surprised many, President Funes himself declined to sign the prohibition into law, instead returning it to the legislative assembly, where it entered a state of legal limbo, neither vetoed nor approved. (In 2016, Funes was embroiled in a corruption scandal that prompted him to flee to Nicaragua, where he was granted asylum. He maintains that he is the victim of political persecution.) The degeneration of the public debate and the failure of the agrochemical ban to become law marked the end of what many had hoped would be a high-water mark for CKDu research and action. After that, in 2014, a new administration took power, and Rodríguez, who was over ninety years old, retired from her role as health minister. Orantes's investigative unit was dissolved, and he was reassigned to a different department. An effort to revive the agrochemical ban foundered, a national action plan that Orantes had developed was cast aside, and the stipulations of the Declaration of San Salvador—which many believed had offered the brightest hope for regional action to combat the epidemic—went largely ignored. It was a "regression that left us in an even worse state than when we started," recalled Orantes.

When his grandfather died of kidney failure in 2005, José López, like most people in Tierra Blanca, had never heard of CKDu. But in the years that followed, *la insuficiencia renal*, as it became known, grew into a dominant feature of life in the Bajo Lempa. Billboards sprung up along the highway connecting the region to San Salvador, advertising private dialysis clinics that few could afford. Local funeral homes began catering almost exclusively to victims of CKDu. ("It's very rare for someone to die from a different disease," one mortician said in an interview with *Undark*.) Nearly everyone was touched, directly or indirectly, by the epidemic. Many field laborers were diagnosed but continued working—compelled to by the grinding poverty endemic to the region. Doctors often dispensed paradoxical advice: Reduce work hours and purchase costly medication.

By the 2010s, the disease was ravaging the López family. One of López's uncles had succumbed in 2009 and now rested beside Juan-Francisco in the Tierra Blanca cemetery. Two more uncles were sick, as was López's father, Vitelio. Worst of all, though, was that both of López's younger brothers had developed the disease— Francisco at the age of ten.

Up to that point, however, López himself had managed to avoid the fate of the other men. His kidneys remained healthy, and he had fallen easily into his role as the family's main breadwinner. Back in 2008, he had married his wife, Marta, and shortly afterwards they had Edwin. López continued to work the fields, and sometimes his ailing father would visit him there, nostalgic for the years they had labored together. The two men would sit together and admire the crops, breathing in the sweet earthy scent of the sprouting sugarcane. "This produce is beautiful," López recalled his father saying. "Perhaps it's even better than when I grew it."

But López's luck eventually ran out. In 2015, blood tests revealed that his creatinine levels were slightly elevated. He showed no symptoms, but doctors nevertheless prescribed medication and told him to reduce his work hours. He tried to follow their orders, but his family's diminishing resources soon made it impossible. Within months, he was back in the fields, working full time.

López's father passed away on April 13, 2019. As with the previous deaths, the family held a two-day vigil and a service with chicken tamales and black coffee. There were religious hymns and a solemn procession to the graveyard. Then Vitelio was laid to rest in the family plot beside his father and brother.

For López's mother, María Luisa, the hardest part of losing her husband was the sense that his death foreshadowed the fate of her boys, all of whom now had CKDu. "There is going to come a day where my kids are going to go," she said one afternoon. "It's a lie to say that they are going to get cured."

Nobody was ever cured. Shortly after Vitelio's death, both of López's remaining uncles followed. There were more tamales and coffee and religious songs. In just over two years the López family plot filled with three fresh graves. And elsewhere in the cemetery, other fresh graves were appearing, too. All across the region, hundreds of families were suffering the same fate.

As the debate over agrochemicals was ramping up in El Salvador in the early 2010s, a competing hypothesis was simultaneously gaining traction within CKDu research circles. Championed by toxic-exposure skeptics like Wesseling, the epidemiologist from the Program on Work and Health in Central America, the theory proposed that heat stress and dehydration, not toxins, were the primary drivers of the CKDu epidemic. The heat stress hypothesis has since garnered widespread support, emerging as the chief competitor to the toxic-exposure theory.

A substantial body of research supports the heat stress hypothesis. In 2012, Wesseling and Sandra Peraza, a chemist at the University of El Salvador, published a study that compared creatinine levels in various communities across the country. Two of the communities cultivated sugarcane as their main economic activity, although one was at sea level while the other was higher in elevation and significantly cooler. The findings echoed the 2005 analysis spearheaded by García-Trabanino, the nephrologist who had first identified the epidemic in the late '90s: Despite employing similar agricultural practices—including the use of pesticides—the high-elevation community had only a 3.6 percent rate of elevated creatinine levels among men. On the rural coast, however, the rate was 28.3 percent. "The major difference," the team wrote, "seems to be ambient temperature in combination with strenuous work."

In 2014, García-Trabanino, Wesseling, and a handful of other researchers conducted a study that further bolstered the heat stress hypothesis. The team tested biomarkers of renal function, such as creatinine and uric acid, in sugarcane workers both before and after a day's shift in the fields. Just one shift, the team found, was enough to substantially increase the prevalence of heightened creatinine levels among the participants, from 20 percent to 25 percent. Similarly, the prevalence of elevated uric acid rose from 26 percent to 43 percent over the course of a shift. Furthermore, the researchers found an association between heat and creatinine levels, observing a 2 percent increase in creatinine for every one-degree increase in temperature. The study provided some of the first detailed data linking heat stress to decreased renal function and stated in its conclusion that "work practices must be improved with more frequent breaks, access to shade during breaks, larger intake of water, and probably also salt."

Based on findings such as these, La Isla Network, or LIN, a nongovernmental organization that focuses on CKDu research and advocacy, began designing and implementing intervention studies in 2015 at El Ángel sugar mill in El Salvador. By introducing a simple program they called "water, rest, and shade" (provision of clean drinking water and electrolyte powder, enforcement of frequent and mandatory work breaks, and the offering of a mobile shade structure) the LIN researchers hoped to do two things: confirm that certain elements of the heat stress theory were correct and determine whether the interventions could reduce workers' risk.

In 2018, the LIN team, which had by then hired Wesseling as chief epidemiologist, published its results in the *Scandinavian Journal of Work, Environment & Health*, a highly cited international occupational health journal. The paper observed two groups of workers, one on the coast and one farther inland. A measurement of kidney function based on creatinine levels was calculated for each worker at the beginning of the harvest, as well as just before the interventions began and again at the end of the season. The interventions, which were implemented two months into the five-month cane cutting season, were shown to slow the rate of kidney decline over the course of the harvest. Amongst the inland group, kidney decline leveled out completely following the intervention. Although the sample size was small, owing to logistical difficulties, with just forty people in the coastal group and forty in the inland group, the researchers were encouraged by what they claimed was the first objective data set demonstrating the efficacy of LIN's occupational interventions.

"In El Salvador, we got the first evidence that heat stress as such can produce incident kidney injury during the harvest," said Wesseling, talking over Zoom from her home in Costa Rica. Subsequent studies at Ingenio San Antonio, a large sugar mill in Nicaragua, have provided further evidence that the interventions are effective—and that the heat stress hypothesis is strong.

Although Wesseling doesn't discount the possibility of other contributing factors, such as toxins, she nevertheless believes that heat stress and dehydration alone are sufficient to explain the epidemic. She and her colleagues at La Isla Network have even developed a hypothesis for a heat-related pathogenesis of CKDu, describing in a 2020 paper how "the release of pro-inflammatory substances from a leaky gut and/or injured muscle" could induce kidney inflammation, which is associated with kidney injury.

Other researchers aren't so convinced. Despite widespread acceptance that heat stress may play a role in the disease's advancement, Marc E. De Broe, the former head of the department of nephrology at the University of Antwerp in Belgium, and Channa Jayasumana, the former health minister of Sri Lanka, believe that something else must be causing the disease to take root in the first place. How could heat alone, they argue, explain the sudden emergence of CKDu in the 1990s, when temperatures in Central America and Sri Lanka were relatively stable over the twentieth century? Or the patchwork distribution of CKDu-endemic areas in Sri Lanka, despite a fairly uniform climate and similar work practices? Or the existence of kidney-injury biomarkers in Central American adolescents who had never worked? Wesseling argues against many of these points on the basis of insufficient or flawed research, yet many researchers remain unconvinced that heat alone can explain the totality of the epidemic's features. As

García-Trabanino put it: "Heat stress seems to be the trigger. But the bullet is already loaded."

A recent study from Sri Lanka seemed to confirm the doubts of many heat stress skeptics. A team from the University of Ruhana compared three occupational groups—rice paddy farmers, tea-plantation workers, and fishers—who labored under varying levels of heat exposure. The team found that the group that worked the longest hours in the hottest weather, fishers, demonstrated the lowest CKDu burden, at 5.36 percent. Meanwhile, paddy farmers, who labored at comparatively low to moderate temperatures, reported a higher incidence of susceptibility, at 13.33 percent. "Our findings indicate that heat stress and dehydration are unlikely to be the leading drivers of CKDu in Sri Lanka," the study concluded, adding, "Heat exposure may act synergistically with other risk factors in causation and progression of CKDu."

The scientific uncertainty has spurred researchers to continue the hunt for García-Trabanino's elusive "bullet." Richard Johnson, a nephrologist at the University of Colorado, is investigating the potentially toxic impact of silica, which is released when sugarcane fields are burned before the harvest. Sandra Peraza, an advocate of heat stress as a major driver of CKDu, is nevertheless curious about the role that genetics may play in predisposing certain populations to the illness. In Sri Lanka, however, the majority of research has focused on toxins, and how various toxic substances may interact with each other to produce the disease.

That each of the two major hypotheses—heat stress and toxins—can be so easily associated with a particular geographical hot spot raises a simple but crucially important question: Are researchers in Central America and South Asia even studying the same disease?

Most believe so, but it cannot yet be definitively proven. One shortcoming that has dogged CKDu research over the years is the absence of a universally accepted definition of the disease. CKDu patients are generally diagnosed clinically, when they present with indicators such as elevated creatinine levels or symptoms of chronic kidney disease in the absence of classic risk factors such as diabetes or hypertension. Pathologically, the disease is generally described as tubulointerstitial nephritis, meaning it causes damage to a certain part of the kidney's filtering units, the tubules, as well as the tissue that surrounds them. But tubulointerstitial nephritis is too common a type of kidney damage to be useful as a diagnostic criterion for CKDu. Still, a 2018 biopsy study led by Julia Wijkström, a Swedish renal pathologist who works closely with LIN, concluded that there were "many similarities in the biochemical and morphological profile of the CKDu endemics in Central America and Sri Lanka, supporting a common etiology." Wijkström's team, however, also noted some differences, and suggested that larger biopsy studies were needed.

In an effort to gain a better understanding of the disease's pathology, a team in 2018 led by Marc E. De Broe (and that included Orantes, the nephrologist who had led El Salvador's CKDu investigative unit before the pesticide debate of 2013) conducted a kidney biopsy study that resulted in a discovery that intrigued the scientists. The cells of all complex organisms contain tiny garbage collectors called lysosomes, which are responsible for breaking down and eliminating waste from the cell. In the specific kidney cells targeted by CKDu, those lysosomes are supposed to be round or oval-shaped, and of a certain minuscule size. De Broe's team, however, found that in CKDu patients, the lysosomes of those cells were enlarged and oddly shaped, like amoebas. That type of lysosomal disfigurement was commonly seen in another group of patients—those who had ingested known nephrotoxic medications. The team then conducted a rat study where they were able to re-create the same lysosomal abnormalities by feeding the rats a drug that is toxic to kidneys called cyclosporine. Another group of rats who were not fed cyclosporine—but who were dehydrated—did not develop the misshapen lysosomes. It was enough for the team to conclude that they had found hard evidence that toxins could cause CKDu.

Additionally, within the aberrant lysosomes the team observed mysterious clumps—or aggregates, as the scientists called them—of an unknown substance, visible under electron microscopy. Cynthia Nast, a renal pathologist who worked on the study, said that the aggregates could be a number of things—the remnants of damaged cell proteins, for example, or the breakdown product of a toxic substance. They might also be, however, the toxic substance itself—minuscule accumulations of whatever agent is causing the disease. But Nast cautions against getting ahead of the science. "Can it maybe at some point inform about causality? Maybe. But we don't know that yet," she said. In the meantime, Nast remains satisfied that her team's research has provided the most compelling evidence yet for a toxic etiology. "I am completely a believer that this lesion is a marker of a toxic nephropathy," she said.

But the study, which was published in the prestigious nephrology journal *Kidney International*, proved controversial. Wijkström, the Swedish pathologist, and a handful of other researchers published a letter to the editor of the journal that claimed they had found the same aberrant lysosomes in five randomly selected healthy kidney donors. In response to the letter, the authors of the original paper replied that the scientists had misunderstood the diagnostic criteria for the lysosomes, and that the lysosomes they had found were not the same as the ones described in the original paper. Within the CKDu research community, attitudes toward the *Kidney International* paper seemed to follow the general contour of the wider CKDu debate: The heat stress people, like Wesseling, viewed Wijkström's letter as a convincing debunking of the lysosomal findings. Meanwhile, the toxic-exposure supporters, such as Nast, viewed the author's reply to Wijkström and the others as proof that their claims had little merit. To this day, even the editor of *Kidney International*, Pierre Ronco, remains torn. "The arguments are well balanced, and the question of the cause remains quite open," he wrote in an email to *Undark*.

If there's anyone who can be said to straddle the divide between the heat stress camp and the toxic-exposure camp, it might be Johnson, the University of Colorado nephrologist. Johnson worked closely with LIN and the heat stress advocates in the first part of the decade, but his research has now shifted toward toxins. For Johnson, however, neither side has decisively proven its case. "If you hear anyone who says it has got to be a toxin, or it has got to be heat stress," he said during a call in July 2022, "you have to realize that they've been convinced when we haven't totally proven it." Johnson acknowledges the significant contributions of scientists from each side of the debate and says that there simply has to be more research. "The lysosomal changes favor a toxin, but there's a lot that favors heat stress," he said. "You keep going until you have absolutely definitive evidence."

And so, the researchers keep going. La Isla Network plans to expand its occupational interventions to more workers through an initiative called Prevention, Resilience, Efficiency, and Protection, or PREP 4 Change. The Belgian team that first identified the strange lysosomes is conducting further research to buttress their claims. And perhaps most significantly, in 2020 the US National Institutes of Health announced a $4 million grant for the future study of CKDu. The project, called the CURE consortium, will be by far the largest effort yet to understand the disease, and researchers from both sides of the debate are hopeful for what it may discover. "This one is getting big and really interesting," said García-Trabanino. "We hope to finally untangle the mystery and stop the massacre."

Just beyond the western edge of town, tucked serenely within a dense grove of mango, cashew, and guanacaste trees, lies the Tierra Blanca cemetery. Its brightly painted tombs suit the wild tangle of the encroaching flora, like enormous tropical flowers. Overhead, the canopy teems with squawking songbirds, but otherwise the cemetery is still. A two-track road weaves delicately through the headstones, just wide enough for a car to pass, and near the far southern edge, beneath the spindly bough of a Morro tree, is the López family plot with its six modest graves.

On a bright afternoon in late March 2022, López's son, Edwin, and mother, María Luisa, headed to the cemetery. The number of graves in the López plot had doubled in the previous three years, and their grief was still raw. When María Luisa reached the light blue cement tomb of her husband, Vitelio, she fell silent, her eyes welling with tears. "All I remember is the sadness," she said, recalling the day the family buried Vitelio. She scanned the six graves with a look of disbelief. "It's not easy knowing that your entire family died from the same disease," she said. "It's like a chain that's tied to them."

Edwin kneeled beside his grandfather's grave and began to sweep it clean with a fallen tree branch. The deaths of his family members have left him feeling abandoned, he explained. Those who remain are forced to struggle forward with whatever resources are left, and must grow up and live without parents, or uncles, or grandparents. "Almost all of them are dead," he said. But for him, the worst has not yet happened. "What I'm really scared of is one day losing my father."

López, who was far too ill to accompany the others to the cemetery that afternoon, shares his son's fear. "I am scared, I will tell you that," he said, speaking of the prospect of leaving his family behind. To temper the worry, López has again turned to his faith.

But he no longer asks for miracles. Now his prayers are only for his family. "I just ask God that whenever the day comes, he gives them strength. He gives them strength to persevere."

For those left behind, however, perseverance is more than a matter of faith. Edwin tries to channel his fear and sadness into motivation for his studies, because it's something tangible he can do to resist the deadly march of the disease. As his father often tells him, his education may be the only way to escape their family's generational cycle of poverty, illness, and death. But it can be hard to concentrate when everything seems so precarious. Francisco, Edwin's uncle, may soon begin dialysis too, and with each of the family's income-earners falling ill, Edwin's chances of avoiding the fields grows slimmer. And regardless of the physiological cause of CKDu, he understands the implications of field work here, in this place. "Agriculture kills you," he said. "That's where the kidney disease comes from."

The Climate Underground

FROM *High Country News*

THE ENTRANCE TO Titan Cave, east of Cody, Wyoming, is hidden on a wide plateau of sagebrush and juniper surrounded by ridge after ridge of craggy mountains. The distant peaks were snowy when I visited in late May, and a slight breeze stirred the desert air. I was with a group of five scientists whose research would take them underground into a grand chamber of stalagmites and stalactites, or speleothems, formations created by occasional drips of water starting hundreds of thousands of years ago or more. They fill Titan's main room with delicate flutes and hulking, lopsided formations that look like something from a seafloor. Hundreds of broken pieces lie scattered around the cave, like piles of bones, while others stand tall, rough stone pillars connecting the floor to the ceiling.

The night before our descent, Jessica Oster, an associate professor of earth and environmental sciences at Vanderbilt University, and one of her graduate students huddled around a laptop open on the bed of the student's motel room in Cody, trying to recall the route to Titan's location on a Bureau of Land Management parcel. Oster, kneeling in front of the computer, sighed. "I'm less worried about this part and more worried about the door," she said, anxiety bringing a lilt to her voice. "I just want everyone to have fun." After a moment, she added, "And stay alive."

The scientists had visited the cave before, but never without a BLM employee guarding its entrance. The door is a heavy metal panel, a couple of feet across, that's supposed to be kept locked. But the BLM cave coordinator would be at an all-day helicopter training, so he'd dropped off a key to the door—along with a sledgehammer. We were on our own.

Lilacs were just starting to bloom in the small towns we drove through on our way to Titan. The scientists pointed out different rock layers through the windows: red siltstone and shale, with names like the Chugwater and Goose Egg formations. Eventually, we reached the top of the plateau, and parked a few yards from the cave mouth.

The researchers stepped around their vehicle and each other, packing up gear, pulling on boots, duct-taping headlamps to helmets. Anticipation combined with the knowledge that we weren't supposed to pee underground meant we took turns ducking behind the scrubby bushes. Earlier, Cameron de Wet, a graduate student, had printed out tiny paper maps of the cave for each of us. Now he carefully adjusted items in one of two blocky blue rectangular bags that held the pieces of a scientific instrument—the reason for the trip.

One of the scientists had analyzed calcium carbonate formations from Titan Cave—stalagmites, the pillars that grow from cave floors—and found that some were around four hundred thousand years old or older. Stalagmites accumulate from the bottom up, preserving the chemical composition of the water that forms them as it drips from the cave ceiling, often from the tip of what looks like a stone icicle—a stalactite. Researchers can use those chemical recordings to infer what the climate was like when the stalagmites formed. But working all this out is complex, and requires understanding the present-day chemical relationships among rainfall on the surface, the water that drips from a cave's ceiling, and the stalagmites below.

The researchers were there to set up equipment to make this easier: an autosampler, an instrument that can be positioned beneath a drip to collect water as it falls. The trip was part of a larger project to help scientists understand what the climate of the Western US was like more than a hundred thousand years ago, using the natural archives of stalagmites and lake sediments.

But first, Oster and her team had to get the entire ottoman-sized autosampler apparatus—clear plastic boxes housing vials and a rotating carousel that holds them, tubing, a funnel, and an expandable tripod to hold the funnel up—deep into the cave, where most of its drips and stalagmites and stalactites are located. There were several obstacles in the way. First was the door, which had a reputation for being stubborn; then a narrow, rocky chute inside the cave's entrance nicknamed "Mr. Twister," which might prove too tight for the autosampler; and then a crawl through a space no more than a foot high. Still, it could be worse, Oster told me, since the crawl was several yards wide—not so narrow that it felt laterally confining. "It's more like being crushed by an anvil," she said.

Oster, de Wet, and another graduate student, Bryce Belanger, walked over to Titan's entrance. The slanted metal door was set into the bottom of a depression nearly invisible behind a small rise. Loose pale rocks lined the short slope down into it; the depression itself was protected by overhanging bedrock and roomy enough for a couple of people to squat inside. The air within was damp and cool, moss blanketing some spots; it felt like a small oasis in the desert landscape.

The scientists scooted down into the depression, then dug out the dirt that had collected along the bottom of the door. They'd been to Titan twice before—in October 2019 and again in September 2021—and once it took them two hours to get inside. Those two hours, however, yielded a crucial insight, which Belanger made use of now: He kicked the door.

That tweaked it enough for de Wet to unlock it. "Whoa," he said as it swung open. "Didn't even need the sledge."

Belanger slid through the open doorway feet-first, into the top of a nearly vertical culvert a couple of feet wide, lined with sturdy rebar rungs. De Wet locked the deadbolt open, so the door couldn't close all the way, and then shut Belanger inside.

Belanger tested the door from below. He pushed it open using both palms and popped up from the hole in the earth, mugging for a camera in the dim light of a cell phone—"That works!"

He turned back around and the rest of us followed him one at a time, our breath loud in our ears in the narrow, echoing culvert. We climbed down about ten or fifteen feet, dropped another foot or two, and then were inside the cave proper. We turned to go deeper in, the wall and ceiling to our left merging into a single diagonal rock face hanging over us. The cobbled slope under our feet was broken by patches of bedrock. We picked our way horizontally across it, following a faint path lit by our headlamps. It had taken months of preparation to get here, and we were excited to finally be underground.

It took only a minute or two to reach the top of Mr. Twister. De Wet disappeared into the chute, shoving one of the blue bags down before him. It's a tricky channel about twenty feet long, with an especially tight spot about halfway down where we had to twist our bodies at the waist so that our hips could squeeze through. Belanger started to feed the second bag down to de Wet, invisible at the bottom of the chute, and the sound of the stiff fabric catching on the rocks filled the cave for a moment. "I got it," de Wet called up.

With the bags safely through, the rest of us followed, sliding and turning our way down Mr. Twister, one by one.

Titan Cave is about a hundred miles from Yellowstone National Park, where, a couple of weeks after our expedition, rain and snowmelt inundated the landscape. Rivers and tributaries demolished previous high-water records; one spot on the Yellowstone— the site the US Geological Survey calls Yellowstone River at Corwin Springs—peaked at 13.88 feet, more than two feet higher than the previous record, set in 1918. The flooding decimated roads and bridges, washed buildings into rivers and broke water mains. The National Park Service temporarily closed the park and ordered more than ten thousand visitors to leave.

Climate change is intensifying weather: Dry periods are drier, wet periods are wetter, and human infrastructure and communities are not, in most places, prepared for it. By the end of this century, the area around Yellowstone is expected to be more than 5 degrees Fahrenheit warmer than it was between 1986 and 2005, and to experience 9 percent more precipitation—but lose 40 percent of the average snowpack. That means more rain, and more floods.

Scientists make these projections using climate models. The models are based on physics: For example, warmer air can hold more moisture than cooler air. This is one reason why, as climate change ratchets up temperatures, storms are becoming more extreme.

Researchers can use information about the past—paleoclimate data—to test how well the models are working. This gives them more confidence in their projections: They can plug certain conditions into the models—like how much of the earth was covered by glaciers or ice sheets, the sea level, the amount of CO_2 in the atmosphere—and then see whether the models return temperature and rainfall patterns that match what actually happened.

"We're pinning our faith on these models to deliver accurate projections," Kim Cobb, a climate scientist and the director of Brown University's Institute at Brown for Environment and Society, said. "And this is one of the most important ways we have of understanding their limitations and their strengths."

This testing, however, requires knowledge of what happened in the past. There are two kinds of historical data: instrumental data and proxy data. Instrumental data comes from direct measurements made with a thermometer, a rain gauge, or another instrument. But the era of direct measurements is just a tiny blip in Earth's 4.5-billion-year history. Oster and her colleagues are particularly interested in the Last Interglacial Period, about 129,000 to 116,000 years ago. Back then, the planet may have been slightly warmer than it is today, similar to the low end of the temperature range predicted for the end of this century.

That could make it a good analogue for the coming decades. It also highlights what Cobb called the most important reason to study paleoclimate records: They can reveal the extraordinary nature of human-caused changes to Earth's climate. The knowledge that global temperatures have not been this high in at least 125,000 years is powerful. "Being able to deliver numbers like that . . . put[s] into full, unfortunately jaw-dropping, context exactly what we're doing right now," Cobb told me.

To understand that context, or as much of it as possible, scientists must employ proxy measurements, like those made from tree rings. But wood rots; even the oldest tree ring data in the Northern Hemisphere goes back only about fourteen thousand years, and it's more commonly used to understand just the last thousand or so years. But other archives last longer: ocean and lake sediments, for example, and cave formations.

For paleoclimate records to be useful, scientists need to know the age of whatever they're analyzing. And speleothems can be precisely dated, said Kathleen Johnson, a geochemist and paleoclimatologist at the University of California, Irvine, and a member of the Grand Traverse Band of Ottawa and Chippewa Indians. The dating method most scientists use, called uranium-thorium dating, is accurate for about the last half a million years, so that's how far back speleothem records typically go—if researchers can find the right ones.

The trouble is that it's impossible to distinguish, from the outside, a 3,000-year-old speleothem from a 300,000-year-old speleothem. Researchers have to crack them open and analyze them to find out. Still, there are some helpful signs: Stalagmites tend to generate a more useful record than stalactites, for example, because they grow in a more straightforward pattern. And a candlestick shape is a good indication of a slow and steady drip rate over time, which makes for a better analysis.

Some people have even developed tricks for finding good samples, Johnson told me, such as shining a flashlight on a stalagmite to see if it lights up like a Himalayan salt lamp—a potential indication of useful calcite—or striking a speleothem and guessing its density from the ringing tone it produces. "I don't think any of them are guaranteed," she said. "But they're, you know, fun to try."

Whenever they can, researchers prefer to take stalagmites that have already broken off on their own, for conservation's sake. Once they select a specimen, they bring it into their lab, then saw it in half vertically, revealing the layers that formed as it grew. Oster showed me a picture on her phone of a cross section of a stalagmite from Titan. They'd nicknamed it "Wee Titan"—it was just under two inches tall—and its layers resembled the strata in a perfectly laminated breakfast pastry.

There are several ways to analyze the layers. One of the most common involves measuring their oxygen isotope signals. These may reflect both temperature and wetness; in the Western US, a higher value may mean colder and wetter conditions, and a lower value may indicate warmer and drier, though some caves show a different pattern. A detailed understanding of how something like rainfall is recorded in the stone of a specific stalagmite requires understanding its context.

Titan Cave, for example, is in an arid location, with certain plants growing above it, a certain soil thickness, a certain kind of rock. "All of that stuff will give it its own personality," Oster explained. Cobb likened this to each cave speaking its own language. Comparing the chemistry of drip water to the chemistry of stalagmites in a cave across several seasons, years, or El Niño/La Niña cycles can help researchers create a sort of Rosetta stone: Once they're able to read which conditions lead to which readings, "We can extend that understanding back through time," Cobb said.

The Yellowstone-area flooding could help the researchers decipher Titan's language. It's unlikely that the cave flooded—it's relatively dry in general—but they wondered whether they might see the heavy rainfall reflected in the oxygen isotope signals of the drip water samples they hoped to collect. If they do, they could apply that knowledge to the older, untranslated layers in Titan speleothems. First, however, they needed to get the water.

At the bottom of Mr. Twister, The cave opened up and we stood on a soft floor of fine, dry dirt, perhaps twenty feet wide, bisected by a path marked by metal reflectors. The reflectors were laid down by the independent cavers who discovered Titan in the late 1980s. There's no natural opening to the cave; according to the BLM cave specialist, the entrance we'd used was dug by cavers, on the advice of a geologist who had felt air coming through cracks in the ground. After that, the BLM installed the door and culvert and closed Titan to recreational caving in order to preserve it for scientific research. Scientists and BLM employees enter occasionally, but aside from some unpublished radon testing, Oster's project was the first to make use of the cave.

De Wet and Oster pointed out a drip site: To the side of the pathway, a bit of the ceiling glistened with moisture. It wasn't forming a speleothem, though, just a small puddle on the floor. They discussed putting out a bottle to sample the drip water, but decided against it.

Something about the close, humid air of the cave made everyone whisper; no one wanted to disturb the subterranean quiet. But just as their hushed discussion ended, we heard an unmistakable plop: a single drop of water falling from the ceiling to the puddle below. "Let's just do it," Oster said.

They put out a plastic bottle a little bigger than a film canister, and de Wet sat to take notes, dust swirling in their headlamp beams. De Wet leaned over the bottle; there was already a drop inside. "Oh, we're in! Great."

We walked deeper into the cave. The next obstacle, the crawl, began gradually: At first, we strolled single file to avoid disturbing the occasional piles of small bones or drip sites next to the path. Then we were crouching, then crawling on hands and knees, and finally inchworming forward on our bellies, our heads tilted to the side so our helmets would fit through, and our feet turned out to avoid the unpleasant sensation of a boot heel catching on the ceiling. Even with the protection of gloves and kneepads, it was tough going, each bit of forward progress the painstaking result of pulling with our fingertips or pushing with our toes, leveraging whatever body parts we could to wriggle onward.

Partway through, the path took a sharp left turn, and then kept going. And going. In the tightest places, previous visitors' passage had compacted the dirt floor, but it was easy to imagine losing the trail and unintentionally shoving my body into an even narrower spot, then being so disoriented I wouldn't be able to find my way out. I tried not to think about the many tons of rock above us.

Finally, suddenly, we were out, in a big open space that felt cavernous after the crawl. "That's about three times as long as would be ideal," de Wet said. "And three times as long as in my memory," Oster replied. But they had managed to drag the blue bags through.

"I can't believe it," Belanger said.

"They better work," Oster added.

We continued, pausing to place the occasional bottle or peer into a sideroom, and taking some time to scramble down a steep cliff, twenty or thirty feet high, with a serendipitous shelf halfway down. Then we were at the entrance to our destination: a large cul-de-sac at the end of one of Titan's passages, an area called the Pisa Room after a prominent column leaning in the middle of it, one of thousands and thousands of stalactites and stalagmites growing from the ceiling and the floor. After the relatively smooth surfaces of the rest of the cave, the Pisa Room was an extravagant profusion of speleothems and wet spots, the air punctuated by an audible drip every few seconds. Many of the formations looked wet, and the stone they were made of was a distinctive milky yellow, reminiscent of mucus. "I feel like I'm inside somebody's nose," Oster said. "Someone with a bad infection."

The researchers launched into action. They downloaded data from instruments they'd left on their last visit—things like plates set underneath drips that count the number of drops over a certain period of time—placed bottles to collect other drops, and evaluated broken speleothems that they might want to carry back out. One of them showed me a helictite, a strange curling thread twisting a few gravity-defying inches out of the side of a stalactite; it's not clear exactly how they form.

Belanger set to work pulling the autosampler pieces out of the blue bags, assembling the instrument and carefully setting the numbered vials—fifty-eight of them, meticulously labeled in the hotel room the night before—in order in the carousel. The carousel rotates, so that a new vial moves beneath the drip every few days; this allows the scientists to analyze how the drip water changes over time. The stalactite that Belanger situated the funnel under looked just like a narrow, two-foot-long carrot hanging from the ceiling: symmetrical and yellow, surrounded by shorter and darker stalactites.

The autosampler was a new piece of equipment from a New Zealand–based company, and the researchers were still working out a few potential pitfalls. The plan was to leave it in the cave until some of them return in September to check on it and collect the full vials of drip water. But a lot could go wrong in the meantime. It runs on a bank of AA batteries, for example—but they could fail. The drip water enters a funnel placed just so—but the funnel could fall. From the funnel, the drips run into a tube—but the tube could pop off the bottom of the funnel. The drips are supposed to flow easily down the tube—but they could get hung up on an air bubble. The tube ends in a pair of needles, which puncture the soft rubber stopper on the vial below—but when the carousel is turning to move a new vial into place, a cap could catch on the plastic case above it, preventing the carousel from getting the next vial in the right spot.

Belanger and de Wet decided to check that last problem. Belanger set the carousel to rotate, but it didn't seem to be working. I asked him if it was doing what he told it to do. "Um, not quite," he said, and bent to pop the two halves of the instrument apart, to see what was going wrong.

Paleoclimate proxy data isn't perfect, so it's a good idea to use multiple archives if possible. And Oster and her colleagues want a broader picture of past climate than any single site could provide. So they didn't confine themselves to Titan Cave; they also looked for clues about the past climate in a cave in California, and previously collected lake sediments from Bear Lake, on the Idaho-Utah border, as well as from lakes across the Great Basin. "The lakes and the caves provide this nice complementary check on one another," explained Dan Ibarra, an assistant professor of earth and environmental sciences at Brown University and a co-leader of the overall project; he heads up the lake portion.

Lake sediment is sampled and stored as cores—columns a couple of inches thick that can be several to hundreds of yards long, collected by drilling into a lakebed. Just like speleothems, the sediment includes layers that record chemical conditions. The deeper you go, the older they get. To interpret that information, researchers need to understand the context of the lake system where they were collected—the chemistry of tributaries at different elevations, for example, or of tributaries fed mostly by snow or by rain. So Ibarra and the team, including the Titan Cave researchers, collected present-day water and sediment samples from Bear Lake and its tributaries.

Just before we visited Titan, the group went to Bear Lake. Over two sunny days, we drove from site to site, pausing frequently to consult maps, to determine whether the road they wanted was private, or simply to let some cows go by. The team split into groups to cover more ground. Both days, I ended up with Natasha Sekhon, a postdoctoral researcher at Brown University who is studying hurricanes and flooding in the Philippines using speleothems. We navigated using her phone's GPS, plugged into the car console screen; she had set the map app to French, and whenever we reached a destination—Sekhon had preloaded the sampling sites Ibarra wanted us to visit—it informed us: "*Vous êtes arrivé.*"

The first day, one of our sites was a stream, a couple yards across, that wound prettily through a cattle pasture, the grass on either side of it dotted with cow pats and dandelions. We parked on a red dirt two-track and walked through a patch of unusually tall sagebrush, our soundtrack a mix of mooing cattle and wind rustling the sage. Oster and Sekhon measured the temperature and pH of the stream as two other researchers, Christopher Kinsley and Warren Sharp, another co-leader of the project, started looking for a good sediment sample. Scientists at the Berkeley Geochronology Center, they determine the ages of the stalagmites and lake cores for the project. Kinsley, in Tevas and shorts, stood in the calf-deep water and scooped up a trowel-full of muck from the streambed. He wasn't happy with the result, though, and let it fall back into the creek.

Upstream, Oster drew water into a syringe, then pulled a filter out of her pocket and twisted it onto the end. Sekhon knelt, holding two small plastic vials, one with a bright green cap and the other hot pink. Oster pushed the water through the filter and into the vials, then into a few additional bottles. Back in the lab, the water samples would be analyzed for isotope signals and for their geochemistry: things like the level of magnesium and calcium, elements that make so-called hard water hard.

Meanwhile, Kinsley scooped up yet another bit of sediment. "I'm getting down to this black stuff again," he said. The black stuff was a layer of sediment with a lot of organic material in it, but Kinsley and Sharp were looking for silt and clay. They intended to use it in their analyses of the lake-core ages. Kinsley brought up another trowel-full. "Might be better," he said as the two picked a few pebbles out, determined it would indeed work, then slipped the sediment into a small plastic bag to take home. "Good prospecting, Christopher," Sharp said as he closed the sample bag. "I thought we were skunked."

The next day, I tagged along with Sekhon, Ibarra, and one of Ibarra's graduate students, Cathy Gagnon, as they sampled more sites. In the early afternoon, we stopped at tiny Preacher Creek, northeast of Bear Lake. The trio was efficient and practiced, moving quickly and in coordination: Sekhon and Gagnon slipped through a barbed wire fence and down a small slope to reach the stream, which flowed into a culvert and under the road where Ibarra stood.

At 6,825 feet in elevation, Preacher Creek was the highest spot we visited that day. It flows into the Smiths Fork, a major tributary of the Bear River. During warmer times, like the Last Interglacial Period and today, the Bear River isn't naturally connected to Bear Lake, but during cooler periods it is. The scientists wanted to make sure they understood its chemistry so they could see how the periods of connection might have changed the chemistry of the sediment cores—a piece of context they'd need to interpret the paleoclimate record.

It was sunny and quiet on the road, the only sounds our voices and the rush of the creek through the culvert below. Then, suddenly, we heard an object hit the water. "Oh no!" Sekhon exclaimed; she'd dropped a sample bottle.

Ibarra ran across the road and down the slope on the far side, hoping to catch it as it came through the culvert. At first it seemed like he'd missed it; then it bobbed into sight and he scooped it up. He carried it back up the slope, but instead of giving it to Sekhon, he tossed it into the car and brought her a clean new one, so that the water sample wouldn't be contaminated. It seemed like a lot of trouble to go to for one errant bottle, but, he said, he didn't want to litter.

Back in the Pisa Room, Belanger and de Wet were pressing the rubber vial caps down harder into the vials, hoping that would help the autosampler's carousel spin the way it was designed to do. Science, of course, like any other human endeavor, is subject to an endless stream of mistakes and corrections, misfortunes and moments of serendipity—in other words, life.

De Wet noticed that the vials were threaded on the bottom and realized that they needed to be screwed into place, to pull them low enough to avoid the lid. Belanger's face broke into a wide smile; he stopped just short of smacking himself on the forehead, relieved to know what the problem was and unperturbed that it could be characterized as operator error. "Oh, that's so smart!" he said with a grin. "Un-be-lievable. So smart."

As they screwed in the vials, de Wet asked if a drip had fallen into the funnel yet. "Yeah, look, there's water coming through," Belanger said, pointing to a drop partway down the tube. Once the vials were done, he reassembled the autosampler, then took out his phone to direct the instrument to rotate the carousel again, worried that the motor had been damaged when the caps had caught the lid. "Gettin' on the Wi-Fi?" de Wet joked, at ease now that they'd figured out what was going on. Belanger smiled. To everyone's relief, the carousel revolved as it was supposed to.

"All right," Belanger said, standing up. The sampler was set. "We're live!" Gentle cheers erupted from the group. As we watched, a single drip dropped into the funnel. "Oh, money," Belanger said, with another big smile. "It ran down! *All right*."

Oster glanced over and saw a drop go down the autosampler's tube. "Oh my God!" She laughed. "I love it!" Then she sighed. "Actually, makes me feel pretty good to see that," she said.

She and Sekhon had been looking at different stalagmites, trying to figure out which ones might have formed during the Last Interglacial Period. They already had samples of the younger, yellow stalagmites—the ones that looked like mucus—and the others that were darker and older. "We're chasing this one little interval of time," Oster said, standing over five broken chunks, debating whether they should take more.

A few minutes later, she collected another that looked in-between as far as color went, which might also mean that it was in-between in age. The researchers numbered the samples with a Sharpie, then wrapped them in brown paper packaging that had originally held the autosampler pieces, packing them into the now-empty blue bags.

By then, we'd been underground for about four hours. As we gathered our gear to head back to the surface, Oster and Sekhon knelt to look at the first vial—a drop had made it all the way to the bottom. "Hopefully, when we come back, it's not still just that drop," Oster said. We turned to go, and Belanger, a smile on his face, looked back at the autosampler—the manufacturer calls it a Syp—one last time. "Be good, Mr. Syp!" he said. "Don't move at all."

It took another hour and a half to get out of the cave—it turns out gravity was a big help in shimmying down Mr. Twister, and an equally big impediment on the way back up—but eventually everyone made it to the surface.

It was a brilliant late afternoon, sunny and hot, the fresh scent of sun-warmed juniper a sharp contrast to the damp air of the cave. The researchers chatted and laughed, the group having jelled in the way that comes from accomplishing something hard together. We snapped a few photos, changed into sandals, and happily ripped open the chocolates that de Wet passed around. As we drove back to Cody, Belanger's thoughts jumped forward to autumn. "I'm just going to be holding my breath when I go back to check on it in the fall," he said.

The next day, we drove to Salt Lake City via Yellowstone. Outside Cody, we passed a layered ridge of pale rock, the same color as the walls of Titan, stark against the gray sky. Oster pulled up an app on her phone—Rockd, created by researchers at the University of Wisconsin-Madison—which showed the geologic formations around us. "Madison limestone! This is it, this is our stuff," she said—the same sedimentary rock layer, formed more than three hundred million years ago, in which Titan Cave is located.

Before we got to the park, Oster and de Wet and I talked about how to define the Anthropocene, a discussion Oster sometimes uses as a class exercise. In the early 2000s, chemist Paul Crutzen suggested that we are living in a new geologic epoch, the Anthropocene, characterized by humanity's impacts on the Earth. Despite widespread popular use, the term has not been officially adopted; that would require affirmative decisions by both the International Commission on Stratigraphy, which is considering it, and the organization that oversees the commission, the International Union of Geological Sciences.

In the meantime, the actual start date of the Anthropocene is a matter of debate: Should it be the beginning of the nuclear age? Humanity's adoption of agriculture? The invention of the Haber-Bosch nitrogen-fixing process, which revolutionized food production by allowing for widespread fertilizer manufacturing? Or perhaps it should start with the colonization of North America, visible in some natural records as a sudden explosion of tree growth across the continent due to the genocide of Indigenous peoples. Oster explained that geologists like to mark the start of an epoch with something physical, a visible layer you can actually point to in rock.

When we arrived at the park, the sky was spitting rain. We drove by Yellowstone Lake, slushy but still frozen, and eventually parked at Norris Geyser Basin, a series of hot springs and geysers transected by trails and boardwalks. The basin is otherworldly: a wide plain dotted with bright green algae, milky blue pools, and thermal vents ringed in chalky white material. Steam rose from the water and disappeared into the low clouds overhead as we walked, stopping now and then to read the signs describing the microorganisms and mineral deposits creating the colors. By then a steady cold rain was falling, but the scientists just pulled up the hoods on their jackets and continued along the path.

MAGGIE KOERTH

The Butterfly Effect

FROM FiveThirtyEight

IT USED TO be, if you wanted to see a Poweshiek skipperling butterfly, the thing to do was go out on the prairie and stare into the middle distance, like you were trying to see a sailboat buried in a Magic Eye painting. Just watch, and wait, and they'd appear.

Back then, more than twenty years ago, there were Poweshieks seemingly everywhere you looked in the Upper Midwest and Canada. Little, erratic, flapping things, Poweshieks are all rust and brown fuzz, no bigger than a quarter. They thrived on the endless golden zen gardens of protected tallgrass, but you could also spot the metallic sheen of their wings dancing among the black-eyed Susans in an overgrown railroad siding, or bounding through grassy ditches pinned between the state highway and a soybean field. "All of a sudden, you'd just see this little sparkle shoot across," said Cale Nordmeyer, a butterfly conservation specialist at the Minnesota Zoo. "Nothing else on the prairie does that."

Now, mostly, nothing at *all* does that. The Poweshieks are all but gone, confined to a patchy mange of small, isolated habitats in Michigan and Manitoba. Even there, survival is precarious, and humans have become a part of the species' life cycle. Researchers collect the white butterfly eggs—each tiny and round like the period on the end of a sentence—and hatch them in captivity, where the caterpillars are supplied with their favorite foods and conditions replicating a perfect Northern winter. In spring, the scientists carry cocoons back to the prairie and release the adult butterflies—guests now on grasslands that once belonged to them.

Our hands are all over these butterflies, yet still they slip through our grasp. This July, in sixteen days of searching through the marshy prairie fens of central Michigan, a team of researchers were able to collect four wild females, perhaps some of the last wild members of the species in the whole United States. By the time you're reading this, these Poweshieks will have died—the average adult lives for only a couple weeks. Some of their offspring are growing on a zoo backlot, slivers of green clinging to stalks of grass in plastic pots. But it's anyone's guess whether other caterpillars are out there under the Michigan sky.

The Poweshiek skipperling is just one threatened species in a world where a thousand tiny things are dying every day, an era of mass extinctions that's been creeping up on us for decades. If humans stop trying to keep the Poweshieks alive, even for a year, the species could disappear. If humans don't stop, it might well go extinct anyway. Even successfully staving off the species' demise would likely mean these butterflies limp into the future changed in fundamental ways—living in different places, under different conditions than they did before Western civilization staked a claim to the prairie.

We are skirting tragedy in a time of certain uncertainty. Nobody knows whether the Poweshiek skipperlings have reached their end, or whether they're starting on a new beginning. Probably it's both, at once. But either way, these butterflies represent a moment of reckoning and wrestling, as environmental scientists struggle to find hope in a world where losing battles are the ones most likely to be fought.

The last known US habitat of wild Poweshieks is a series of small prairies, the biggest maybe the size of a city block, clustered around a lake in Michigan. The specific location is kept a secret in order to protect the butterflies from both collectors and well-meaning lookie-loos who might otherwise come out to photograph the Poweshieks and, inadvertently, grind the caterpillars into the damp soil. It's a wet place, full of matted paths that twine through clots of muddy grasses and waving wildflowers. The only noises are grasshoppers, massasauga rattlesnakes, and the wind.

The conditions there weren't perfect for Poweshieks on the warm afternoon in July 2022, when Nordmeyer took me along on his hunt. Clouds were passing overhead, blotting out the sunshine butterflies love, and the breeze was up, maybe enough to make a small, winged creature cling to a flower as if it were a port in a storm. But there were other butterflies about. Movement in the grass stirred mulberry wing skippers and dorcas coppers. They flew into the air, where a hopeful journalist repeatedly mistook them for Poweshieks.

In mud-caked boots, we walked through areas Nordemeyer and his team called Poweshiek Country and Poweshiek City—named less than a decade ago to reflect their relative skipperling populations. But now both felt like ghost towns. Nordmeyer, who has been deeply involved in efforts to boost wild Poweshiek populations in Michigan, wasn't worried at first. He'd come to this place regularly since 2012 and he figured that maybe it was just a little early in the season. It had been a cold spring. Maybe the wild butterflies were delayed in emerging from their cocoons. Or maybe the day was just bad. It seemed hard to believe they were gone entirely.

But that's the same hope researchers have been clinging to since the species started vanishing, more than twenty years ago. Back then, Robert Dana couldn't believe it, either, when he first started hearing that Poweshieks were going missing on the Minnesota prairies. It was the scientific equivalent of rumors. A small survey here. A set of anecdotal reports from hobbyist bug lovers there. Dana, then an entomologist with the Minnesota Department of Natural Resources, didn't buy it. It was like somebody saying the grass was gone. From the first time he'd ever seen a prairie as a graduate student, Poweshieks had been there. It wasn't the biggest butterfly, or the showiest. It was humble but distinct. Beautiful in its own quiet way. A real Upper Midwesterner of an insect.

Dana believed the Poweshieks must be okay in the early 2000s for the same reasons Nordmeyer hoped they're going to be now: It's hard to imagine a world without them. The Poweshieks once defined the Northern prairies, with a range that stretched down from Manitoba, across both Dakotas, Minnesota, Wisconsin, Michigan, and even south into Iowa, Illinois, and Indiana. The butterfly's name even comes from a county in Iowa and, by extension, from an early nineteenth-century chief of the Meskwaki people.

And while the years between Dana's search and Nordmeyer's have seen a massive retreat of Poweshiek territory, it remains difficult to know if any single bad year for the species is truly a step on the road to extinction. Poweshieks are small butterflies, compact and fast—they don't really fly so much as manically hop across the landscape from flower to flower, a dark blur just above the top of the grass. They do not travel long distances in the single year that makes up an individual's lifetime. The tiny caterpillars, nearly invisible on a bending stem, stay within a few centimeters of where they hatch, waiting out the winter beneath the snow. A typical flight for an adult is only a few meters at a time, and it may not leave an area bigger than a square mile before it dies.

Their short lives and small individual home ranges mean that local populations have always fluctuated a lot from year to year, and it's easy to lose a population in one small, specific location while a different population flourishes nearby. This is what Dana assumed was actually going on in Minnesota back in the early 2000s. People who were looking in only one place, one year, were just mistaking these local fluctuations for actual disappearances, he told himself.

Yet Dana found a hollow stillness everywhere he went, like a room grown suddenly too silent. Across more than fifty locations, his 2006 search turned up exactly one butterfly. As the horror of what he was witnessing began to truly set in, he clung to that butterfly as a beacon. "I guess I felt some kind of relief that maybe it was still hanging on, maybe it was going to recover," Dana said. It was the last Poweshiek he'd ever see in the wild.

Nearly twenty years later in Michigan, Nordmeyer would come up with similarly grim statistics. There were no Poweshieks in Poweshiek City. There were almost no Poweshieks anywhere. After I left the prairie patches, Nordmeyer stayed on, and it became clear the absence wasn't just a product of bad timing. But, like Dana before him, Nordmeyer wasn't ready to give up. Instead, he contacted the federal officials in charge of endangered species management and came to an agreement—whatever few wild Poweshieks he found would be taken into captivity and mated with the butterflies he had raised by hand.

"Fish and Wildlife [Service] pretty much decided, 'No, we think these things are safer with you guys than they are in the wild,'" Nordmeyer said.

Today, Poweshiek skipperlings are ferried to adulthood in the back seat of a Subaru Outback. The ride from the Minnesota Zoo to Michigan is the culmination of their new, human-directed life cycle. They make the journey still wrapped in their cocoons, no longer caterpillars but not quite yet butterflies, each one attached to a small tuft of prairie dropseed growing from a plastic pot and wrapped in a protective tower of pantyhose-covered metal framing. Nordmeyer drives the Poweshieks to Michigan, where he sets up the pots like a miniature city of beige condominiums inside a collapsible picnic shelter on site. He calls it the "Poweshiek Party Tent."

As each cocoon opens, the butterfly that emerges is released into the wild. Two or three at a time, Nordmeyer marks their wings with an identifying dot from a colored Sharpie, loads them into test tubes and carries them out onto the prairie in a messenger bag. He lowers them carefully by the bristly end of a paintbrush onto the waiting petals of a black-eyed Susan, their favorite flower. Over the course of a couple of weeks in 2022, Nordmeyer and his colleagues released 102 butterflies and searched the grasses for pregnant females and already laid eggs to carry back to the zoo. The caterpillars that later hatched from those eggs went on to eat their way through their own potted prairie dropseeds. In the summer, they were loaded into special cooling boxes that can mimic the overnight temperature drops that no longer happen reliably in this part of the country. As the seasons turned, humans plucked the fat babies from the plants and packed them into plastic cups filled with clay. The cups are covered with paper towels and stored in a freezer, the analogue of a caterpillar buried beneath a thick snowpack. In the spring, they'll go back on a dropseed, spin a cocoon, and wait for their cross-country road trip in that Subaru—the new year's butterfly crop.

Nordmeyer isn't the only one working to save the Poweshiek. At the John Ball Zoo in Michigan, scientists have figured out how to breed Poweshieks right there in captivity, no need to drop them off in a field so they can find dates. In Manitoba, researchers are monitoring Poweshiek numbers and working to figure out both what kinds of habitats the butterflies prefer and ways of keeping those habitats healthy and safe.

Twenty years ago, when Dana first called Richard Westwood, a professor of biology at the University of Winnipeg, and asked him if he'd noticed anything odd about the Poweshieks in his area, Westwood's response was "I don't know. They're always around right? We don't really pay attention." Today, they are a species

Westwood and dozens of other scientists have spent years paying intense attention to.

The Poweshieks' transition from a species humans neglected to one whose life is now literally in our hands is a reflection of how we've approached insects, as a whole. For every charismatic mega-fauna gracing the cover of a magazine, there are literal armfuls of smaller creatures fading quietly into the night. Insects are threat-ened at a rate far exceeding that of mammals, birds, and reptiles, with as much as 40 percent of all insect species potentially facing extinction in the next few decades. Butterflies are one of the most affected groups, but they weren't a major focus of conservation sci-ence until recently. The International Union for the Conservation of Nature, the agency that evaluates the conservation status of spe-cies and helps determine which are threatened, has assessed the risks posed to 67 percent of vertebrate species but only 2 percent of invertebrates, as of 2019.

Saving those species is an exercise in just how much control we can take over nature before we're not only preserving it but also molding it. The people working to save the Poweshiek know they're walking that tightrope. And they also know it's not some-thing that we have enough resources to do for every species that's in danger of extinction. Choices will have to be made.

That reality represents a big shift in the way both science and policy have thought about animal conservation, said Daniel Rohlf, a professor of wildlife law at Lewis & Clark Law School. There's a can-do attitude that's embedded into the law and regulations that govern how species are managed. The Endangered Species Act de-fines the very concept of "conservation" as doing whatever is nec-essary to ensure every listed species no longer needs the ESA. "The overwhelming attitude was, 'Humans have kind of messed this up, but we can fix it,'" Rohlf said. Species become endangered, but then humans come in, identify what went wrong, correct the prob-lem, and help the species rebuild until eventually it stops needing our intervention.

There have definitely been some species for which that under-standing of the world has worked. Melinda Morgan, director of the sustainability studies program at the University of New Mexico, pointed to the success we've had in saving the peregrine falcon. "Ban DDT and you're great. Done," she said, referring to the once-common bird-killing pesticide. But the reality is that most endangered animals will not be large and iconic, with an easily identified threat that can be quickly eliminated with one weird trick, like in a clickbait ad. Most are insects, small and hard to find, easier to accidentally kill without thinking about it than to find and rescue. Most are suffering from a tangle of intertwined problems. For the Poweshiek, there are pesticides we won't stop using, land we won't stop developing, and climate that won't stop changing. If you resuscitate the species in a zoo, it may or may not have a wild habitat to return to. Most endangered animals are not the peregrine falcon. Most endangered animals are a mess.

"There's a level of overwhelm that comes with that," Morgan said. "There's a level of despair." What do you do when a species you thought was fine turns out to be actually teetering on the brink of death? What do you do when you have to decide which species you'll focus your grants and labs and manpower on, knowing there are others in just as dire straits? What do you do when you have basically become a minor deity to a species of butterfly that relies on you to guide its life across generations—and the damn thing won't be fruitful and multiply?

This is the point where Morgan suggested I speak to a Buddhist philosopher.

Three years ago, researchers from the Minnesota Zoo took one of the Poweshieks they'd painstakingly raised from birth, released it onto the Michigan prairie, and watched in frustration as a mint-green dragonfly, the size of a human palm, dropped out of the sky . . . and ate it.

There are no guarantees in nature. Not even when everything is working the way it's supposed to—maybe especially not then. The dragonfly, a common pondhawk, was also native to those swampy grasslands. "It sort of forced us to stop and think," Nordmeyer said. "That's also a natural species, right? This is part of the natural order of things then, too."

This *Far Side* cartoon of a moment is where the hard science of keeping a species alive runs smack into philosophy. That's where Joanna Macy comes in. Macy is a ninety-three-year-old environmental activist and Buddhist scholar whose work focuses specifically on the kinds of challenges scientists face when they have to decide how far they're willing to go to conserve a species like the Poweshiek skipperling. Her writing is dedicated to staring straight down the barrel of environmental failure and coming away with a heart that's larger, rather than one that's been blasted to bits.

How do you do that? Well, consider the idea that the world is a bit like a tomato. In *Active Hope*, a 2012 book Macy wrote with psychologist Dr. Chris Johnstone, the tomato helps illustrate what human intervention in nature can often look like. If you squeeze a tomato too much, too hard—when humans alter our environment in unsustainable ways—we destroy it. You can't unsqueeze a tomato, just like you can't unmush the world we've damaged. If the chances of fixing the problem aren't very high, then why not just stop trying to help altogether?

But the analogy doesn't end with the world destroyed in a pasta-sauce apocalypse. Instead, Johnstone told me, every collapse carries the seeds of a future renewal. What grows will be a different tomato. You can't get the old one back. But something can grow—if we take the seeds and plant them.

And that's . . . it. That's the message. Somehow, that's supposed to be uplifting? Buck up, little buttercup, and keep trying? It's okay if that's not enough for you. It wasn't enough for me. But then I realized that, whether the scientists I was interviewing knew about Macy and Johnstone's work or not, they had already come to rely on this perspective for their own sanity. They weren't giving up because hope, for them, wasn't dependent on the Poweshieks' odds of survival. Instead, hope was an action to take. They wanted something good to happen, so they tried to make it so.

That's why they could accept the risk of Poweshieks being eaten by natural predators—their efforts don't need to be successful to be worthwhile. If the Michigan summer truly ended without any Poweshiek caterpillars clinging to the drying grasses—if the species is truly gone in the US outside of captivity—it's all still been worth it, said Anna Monfils, a professor of biology at Central Michigan University. That's in part because the struggle is bigger than those prairie fens, and bigger than the Poweshieks. If the skipperlings died there this year, but live on in zoos and in Canada, maybe that can teach us something about what's happening in this specific ecosystem and how it might affect the other animals that live there—the other butterflies, the dragonflies that eat them, or even the rattlesnakes hidden under the grass. "No loss is good," she said. "But learning from that process is a better outcome than not learning."

It matters that researchers can breed Poweshieks in captivity now, for example—not just for the Poweshieks, but for butterflies on the other side of the world who might benefit from the same techniques. And, as multiple scientists pointed out, the only way to really guarantee failure is to stop.

Macy and Johnstone describe this as radical uncertainty, the realization that "we don't know if this will work" goes both ways. It could mean a world where the only Poweshieks are stiff and pinned to cardboard, under glass, a memorial to themselves. But uncertainty can also mean scientists' intervention works, putting the species in a good position to be successfully reintroduced somewhere else. It could mean a young scientist comes along who figures out how to ensure the prairies the Poweshieks once loved are safe for them again. It could mean, as Nordmeyer hoped, that we find "a new pocket of these butterflies that we didn't even know was in someone's backyard." It isn't naive to think things might not be as bad as they seem. And that interpretation provides a reason to do the work.

This winter, the community of scientists who have taken on that work will talk about the next steps to take with the Poweshiek now that the Michigan prairie fens are no longer a place the butterfly seems to thrive. Something will change. What that is, nobody knows yet. It's possible they'll stop trying to release Poweshieks in Michigan at all, and instead focus on breeding for a few years, working up the captive population so it's large enough to try reintroduction at a different site. But it's unlikely this is the end. There's still too much to hope for.

"When you get us all together in the room we can be extremely positive," Westwood said. "It's depressing. But I don't see anybody here giving up. We won't give up until it's give-up time."

Because that's the other thing that seemed truly important to these scientists: the fact that others were there, seeing the things they saw, carrying the burden with them. The ecologist Aldo Leopold once wrote that "one of the penalties of an ecological education is that one lives alone in a world of wounds." But that's not how these researchers live. They're part of a community that crosses borders and has spent years helping a tiny creature rebuilt its own community. They may not succeed. But they won't have been alone. And maybe that, by itself, can heal some wounds.

BEN MAUK

Shadows, Tokens, Spring

FROM *The Virginia Quarterly Review*

I

The Mongolian marmot, or tarbagan, is hunted in the fall, when the animal has prepared for hibernation by fattening itself on the berries, roots, and lichens of the Altai Mountains. It has been hunted for as long as humans have hunted.

A Mongolian dish known as *boodog* calls for fire-heated stones to be placed into the abdomen of a deboned and disemboweled marmot. "Hang a marmot or a goat at the head and cut the skin around its neck," one recipe instructs. "Now it is possible to pull the skin and most of the meat down over the inner skeleton." The chef must break the legs at the knee to remove the femurs—it seems unwise to attempt this dish indoors—and, from the innards, keep the liver and kidneys, which will be reinserted later on. Then comes the dish's signature trick:

> Turn the removed skin and meat back, so that the hair is at the outside again. Fill the resulting "sack" with the following ingredients: Some salt, one or two peeled onions, and a number of stones, that have been heated up in a fire for about an hour. The stones must have a smooth and round surface. The smaller ones go into the upper legs, the larger ones into the abdominal cavity.

If the skin goes tight when cooking, cut a few small holes in the carcass to release pressure. The meat, heated from the inside out, is finished when the skin leaks with fat.

The invention of this method of cooking requires no great imaginative leap. A marmot in the wild already resembles a formless brown bag of flesh, just as in life it seems to contain little but its own coiled panic, a hard motor swaddled in softness that allows the animal, when threatened, to burrow into the soil with electric speed.

Marco Polo may have described the tarbagan in his chapter on the Tatars, who, he claimed, "subsist entirely on flesh and milk . . . and a certain small animal, not unlike a rabbit, called by our people Pharaoh's mice, which, during the summer season, are found in great abundance in the plains." Perhaps because the term otherwise describes a type of mongoose not found in Central Asia, Marco Polo's "Pharaoh's mouse" has sometimes been identified as the tarbagan, whose fecundity in summer is legendary. Although he did not mention either chopsticks or the Great Wall on his journey to the East, the Venetian merchant counted marmots as worthy of his *Book of the Marvels of the World*.

The restless doctor John Bell left a more detailed description in his only book, the two-volume *Travels from St. Petersburg in Russia, to Diverse Parts of Asia:* "On these hills are a great number of animals called marmots, of a brownish color, having feet like a badger; and nearly of the same size. They make deep burrows on the declivities of the hills; and, it is said, that, in winter, they continue in these holes, for a certain time, even without food." The Jesuit priest Jean-Baptiste du Halde recounted the hunting of *un très-grand nombre* of tarbagans, which he described as a species of *rat de terre*.

To watch a marmot in flight is to behold the transubstantiation of unassuming plant matter—clover, stonecrop, moss—into a fur-lined reservoir of energy. As Bell observed, even during their hibernation period marmots "sit or lie near their burrows, keeping a strict watch; and, at the approach of danger, rear themselves upon their hind-feet, giving a loud whistle . . . and then drop into their holes in a moment." I have watched marmots move through forest undergrowth with an elegance and swiftness that rivals the antelope. Once, in the mountains of Kazakhstan, I saw a marmot cross a treeless field and thought at first it was the shadow of a bird.

Marmots are the largest of a diverse and widely distributed (or "cosmopolitan," as taxonomists say) collection of small plant-eating rodents belonging to the family *Sciuridae*. Their lighter cousins include squirrels and chipmunks. There are yellow-bellied marmots and forest-steppe marmots, marmots living high in the Alps, in the Apennines, and in the Rocky Mountains. There are marmots across the American prairie—not to be confused with the smaller and more social prairie dog, another *Sciuridae*—and across the Great Eurasian Steppe that extends from Hungary to the Great Wall that Marco Polo did not describe, where since 2008 the tarbagan has been listed as an endangered species.

In point of fact there are fourteen known species of the *Sciuridae* genus *Marmota*, nine of which are found in Eurasia. Their diversity is the result of transarctic migrations, the most recent of which took place more than a million years ago. In the Altai and Tian Shan ranges are found rose marmots and pink marmots, black-capped marmots and long-tailed marmots. Each embodies the same paradox of size and speed, and the differences between them may seem academic. But one trait, shared among some but not all species, has determined the course of history. The peoples living east of Lake Baikal have long described the "marmot poison" some animals bore inside them. In recent years, the global scientific community has arrived at the same knowledge. They have identified four marmots—the gray, the red, the Himalayan, and the Mongolian (or Siberian) marmot, which is the tarbagan—as bearers of the plague that has wiped out large swaths of humankind.

2

In Richard Kephale's *Medela Pestilentiae* they are called "Gods Tokens": dark lesions on the skin, eruptions of subcutaneous hemorrhaging "the bigness of a flea-bitten spot, sometimes much bigger." They appeared clustered on the backs and breasts of corpses, it was theorized, because "the vital spirits strive to breath[e] out the venom the nearest way." The token's color revealed the victim's predominant humor. A body with red tokens was filled with choler; black, melancholy; blue, phlegm.

Daniel Defoe, who owned a copy of *Medela Pestilentiae*, described the discovery of these casement windows to the soul in some of the horrifying scenes in *A Journal of the Plague Year*. The plague that struck London in 1665, he wrote, "defied all Medicine; the very Physicians were seized with it, with their Preservatives in their Mouths; and Men went about prescribing to others and telling them what to do, till the Tokens were upon them, and they dropt down dead, destroyed by that very Enemy, they directed others to oppose."

Elsewhere in the book, he tells the unforgettable story of a mother and daughter:

> While the Bed was airing, the Mother undressed the young Woman, and just as she was laid down in the Bed, she looking upon her Body with a Candle, immediately discovered the fatal Tokens on the Inside of her Thighs. Her Mother not being able to contain herself, threw down her Candle, and shriekt out in such a frightful Manner, that it was enough to place Horror upon the stoutest Heart in the World; nor was it one Skream, or one Cry, but the Fright having seiz'd her Spirits, she fainted first, then recovered, then ran all over the House, up the Stairs and down the Stairs, like one distracted, and indeed really was distracted, and continued screeching and crying out for several Hours, void of all Sense. . . . As to the young Maiden, she was a dead Corpse from that Moment; for the Gangren which occasions the Spots had spread [over] her whole Body, and she died in less than two Hours: But still the Mother continued crying out, not knowing any Thing more of her Child, several Hours after she was dead. It is so long ago, that I am not certain, but I think the Mother never recover'd, but died in two or three Weeks after.

The scene anticipates by centuries a contemporary movie trope: the zombie bite, wherein a victim—as yet untransformed—reveals the wound that marks her as already undead. That Defoe was only five years old when the plague struck does not mitigate the book's horrors. The pandemic he described feels immediate and contemporary, and probably owes something to the memories of Defoe's uncle, who, like the narrator of the *Journal*, was a saddler with the initials H.F.

All forms of plague are thought to have their evolutionary origins in Inner Mongolia, a "big bang" of multiple genetic lineages, although their epidemiological journey is ancient and obscure. It seems probable that the Altai Mountains are involved. The same coccobacillus, *Yersinia pestis*, which originated in these mountains in some prehistoric and initially harmless form, is responsible for all three known varieties of plague. They are differentiated by the locus of infection. Septicemic plague infects the blood, and thankfully is not itself easily contagious, instead usually accompanying one of the other two types. (It might otherwise constitute the most dangerous variety.) Bubonic plague attacks the lymph nodes. It is transmitted from rodents to fleas and from fleas to humans, but does not pass from one human to another. The Black Death that struck down fifty million people in the 1300s was almost certainly bubonic plague; it killed upward of 60 percent of those it infected in Europe, mostly the poor. The Great Plague of London was similarly bubonic in nature, part of the centuries-long "second pandemic" in Europe. By some estimates, nearly a quarter of the city's population perished. But when that same bacterium *Y. pestis* causes a lung infection, it becomes far more contagious than septicemic and around twice as deadly as bubonic plague. Transmitted via droplets in the air, there is no cure unless the disease is caught at the onset of symptoms. Left untreated for twenty-four hours, the patient is guaranteed to die. This is pneumonic plague.

3

In the depths of the Great Manchurian Plague, which claimed sixty thousand lives between 1910 and 1911, it became impossible to dispose of the proliferating dead. Because homes of the infected were required to undergo police investigation and disinfection at the owners' expense, corpses were tossed into the street surreptitiously at night. Authorities would discover the freshly sprouted bodies in the morning, when there was nothing left to do but place them in coffins and transport them to burial grounds outside town.

But the winter earth, frozen seven feet down in Manchuria, could not be shoveled out, so the coffins were laid out in rows by the side of the road. Sometimes the police didn't find a body right away and it froze in a complicated position, in which cases the thin, unplaned government coffins were left unfastened or even open with frozen arms and legs protruding at odd angles. Sometimes the body had frozen while curled in a fetal position or else seated upright, and remained that way, as though piloting the coffin like a canoe. Eventually, there were no more coffins. The line of corpses stretched on for more than a mile of snow-covered ground.

The doctor who observed this line of unburied dead was a Penang-born British subject named Wu Lien-Teh. Wu and his Cantonese assistant arrived at the train station in Harbin on the bitterly cold afternoon of Christmas Eve in 1910. A droshky led by two Mongolian ponies was waiting for them. The porters who went for his luggage were all Russians, dressed in sheepskin jackets and padded trousers, and they wore stiff, felted knee-high boots. The air was dry and ice crystals formed on Wu's eyebrows. The droshky took them to a hotel run by a "Russian jewess" in the business district, Wu recalled. Her burly assistant, a migrant worker from Shandong, ran the doctor a hot bath.

Wu was then vice director of the army medical college at Tianjin. He'd left Peking on emergency orders from the Ministry of Foreign Affairs and had traveled by train for three days across northern China, past the eastern end of the Great Wall and past Mukden, capital of the Three Eastern Provinces, before arriving at Harbin, a soybean boomtown contested by the great powers. Adjoining Russia-controlled Harbin was Fuchiatien, a town of twenty-four thousand people under Chinese administration, where officials had reported several cases of an unidentified fatal illness. Its symptoms followed the same order each time: fever, coughing, blood-spitting, a purpling of the body's skin, and finally, death.

Most of the early victims were trappers from Manzhouli, a trading post just across the border in Russian territory. It was another frontier town surrounded by wilderness. The trappers, most of them migrant workers from Shandong, traded in the large Mongolian marmot, whose fur was dyed and treated to make imitation sable for undiscerning women in Europe and America. They worked out on the open steppe in conditions of oaken fortitude, carrying their water with them and eating frozen dumplings in the disorientingly dry and frozen grasslands. After collecting two dozen or so tarbagan furs over a period of several days they came back to Manzhouli and slept in the basement rooms of crowded inns, where the air became humid with breath and sweat. Here one man lay down next to another; conditions for the spread of disease were nearly perfect. As soon as they sold their skins, they went out again onto the steppe, like old men at a sauna, alternating between rooms that heat and pools that freeze.

The mysterious coughing and blood-spitting disease had spread slowly in the months before Wu's arrival. The first cases among the trappers frightened people in Manzhouli enough that those with the means to do so fled east and south to Harbin on the Chinese

Eastern, a new single-track railway built by Russia after the first
Sino-Japanese War. The railway was under Russian jurisdiction, but
it crossed plains and forests for more than five hundred sparsely
populated miles of Russian-, Chinese-, and Japanese-controlled territory. All along the route, the infected disembarked and spread the
plague into towns where sanitary conditions were generally loath-
some. Although it stood adjacent to the high stone buildings of
the Russian quarter, Fuchiatien was a shantytown of wooden shacks
and corrugated roofs. Streets were impassable mud-ways flanked by
wooden planks for pedestrians, a Venice of the underworld.

When traders and workmen came to Harbin, they passed
through a train station and storeyard with hundreds of wagons of
soybeans. The wagons were uncovered since it only snowed and
never rained in winter. Along with furs, meat, millet, and timber,
the soybeans were shipped south to Changchun, to Mukden (now
called Shenyang), to the Japanese port at Dalian, and to Russia
by way of Vladivostok. The city was an early outpost of globalized
trade. Harbin was also segregated by nationality and class, with Chinese laborers, called "coolies," in the lowest position. After passing
the storeyards, lower-class workers proceeded to the crowded Chinese town or, if they were in the Russian quarter, chose a cramped
Chinese inn, where they slept on a large common bed called a
kang. Made from bricks covered in blankets, the kang was heated
from below the house by an outdoor fire. Guests slept, sat, dressed,
and ate there, pressed together with the windows shut all winter
long against the cold. Such conditions prevailed in towns and
villages throughout the Three Eastern Provinces.

Early attempts to control the plague were rustic. In his autobiography—a boosterish account of one man's rise to international prominence and of the triumph of modern medicine over
primitive superstition—Wu catalogues these efforts with irritation.
A plague house was set up in a former public bathhouse, where
men with fever, spitting blood, were taken to die. But there was no
attempt to isolate people who found themselves in close contact
with the dead. The European doctors working in Harbin were convinced that rats were the main vectors of the disease, so a futile rat-
killing operation was mounted. Two doctors named Yao and Sun
rented a mule-cart depot and filled carts with sulfur and carbolic
acid, traditionally used to rid houses of demons. Neither substance
managed to slow the spread of disease, but *shih t'an suan*, one of

the names for carbolic acid, filled the houses of the bereaved with the consoling stench of sanitation.

Wu was there to impose a modern approach. He had studied at Cambridge and spoke with officials in broken Mandarin Chinese and fluent English, finding the latter especially useful for medical terms. He did not make much use of his German or French, which he'd picked up in those countries in the course of his bacteriological research. He found all racial distinctions absurd and offensive, claimed never to have experienced any discrimination while in England, and relished his role as a man of science who believed in the free transmission of knowledge across borders and among men of all races.

On December 25, the day after his arrival, Wu learned that ten deaths had been reported to officials in the morning. (Before long, as many as fifty were dying each day.) On December 27, he performed an autopsy—he called it the first-ever postmortem of a victim of the Black Death in Fuchiatien, and perhaps in all of Manchuria—on a Japanese innkeeper who had died overnight. She was laid out on a soiled tatami atop some raised wooden planks, wearing a cheap cotton-padded kimono. Wu sent for water and made his incisions in the woman's dirty home. There was no laboratory for this kind of work.

Wu's description of the event is clinical. "After the cartilaginous portion of the chest had been removed, a thick-bored syringe needle was plunged into the right auricle and sufficient blood was removed for culture in two agar tubes and for thin films on slides . . . and a platinum needle was inserted into the substance of each organ and the necessary cultures and films made. Pieces of the affected lungs, spleen, and liver, each two inches by two inches, were removed and placed in glass jars." The skin was then sewn up—the autopsy was kept secret from the innkeeper's family—and the body taken to the burial ground. Back at his hotel, Wu managed to confirm, with a simple staining of Loeffler's methylene blue under his travel microscope, that the woman's body was teeming with *Y. pestis*.

After the autopsy, a laboratory room was set up in the disinfecting station where Wu kept his slides and cultures, and where he often consulted a Japanese doctor who had been sent by the South Manchuria Railway. Wu tried to convince the doctor that the disease "was spread principally by direct coughing of dangerous bacilli in the sputum expelled from diseased lungs" and that household rats were not the culprit. The Japanese doctor could not be swayed. Others thought Wu ridiculous; the European consuls treated him superciliously or, at best, as a curiosity. The French vice-consul was briefly impressed when Wu mentioned his studies at the Institut Pasteur in Paris, but his credentials did not change the prevailing wisdom among doctors in Manchuria: that the plague could not be transmitted directly between humans. Another Frenchman, a doctor named Mesny, considered the theory of rat-borne transmission his personal property. He raged at Wu for his suggestion that it was wrong. In Wu's telling, at one meeting he raised his arms "and with bulging eyes cried out, 'You, you Chinaman, how dare you laugh at me and contradict your superior?'" Nine days after their meeting, his skin purple and covered in buboes, Mesny died.

If Wu's heroic account is to be believed, Mesny's death was an inflection point in the crisis. Chinese authorities began to treat the plague as the epidemic Wu believed it was, and to respect his authority in the fight to contain it. His theories of aerosolized transmission gained traction. Medical workers, disinfectors, and gravediggers were advised to wear masks, either a wire-meshed frame covered in black muslin or a soft piece of surgical gauze covered in cotton. In both cases, Wu often saw the masks hanging loosely from the necks of their wearers. He made a note to design a comfortable anti-contagion mask for workers once the emergency was over.

Chinese migrant workers still considered the plague a shameful disease; they asked doctors not to reveal their cause of death to relatives. Shandong, a great exporter of migrant labor, was filled with households whose members never learned why a son or brother had vanished in Manchuria.

Workers also resented the cruelty of the Russian police, who rounded up "coolies" indiscriminately as plague suspects and locked them together in long rows of railway cars. There were rumors that Russians forced Chinese to undress in the open field outside the quarantine wagons and that some died from exposure. Sick family members fled their homes to avoid endangering relatives. The Russian "flying squad" hunted down the dying to bring them to the plague hospital or wagons from which no one ever emerged. One newspaper wrote that Russians were using the plague as a pretext to banish the Chinese from town centers altogether. Among Russians, it was thought that the Chinese government's laxity was the reason the plague had reached Harbin in the first place. The railway stopped selling third- and fourth-class tickets in order to block the mass movement of laborers. (Upper-class citizens were exempt from all movement restrictions.) The ticketless moved south by cart or by foot; the fallen soon lined the roads leading out from Harbin. Others died of cold and hunger after being refused entry into villages along the way.

After visiting the frozen burial grounds in January, Wu began to advocate for a campaign of mass cremation. He invited local officials to drive out to see the coffins and corpses; there were two thousand unburied plague victims sitting in the snowbanks, their bodies wrapped in cloth soaked in carbolic acid. (The coffin shortage was partly due to rampant corruption among Chinese officials; plague relief funds were easy to plunder.) Wu feared that, although they were not the primary vector, rats would gnaw the corpses and spread the disease even further.

Cremation was a sacrilege. It took a formal sanction from the Qing emperor for the mass burning of corpses in "plague pits" to begin. Poor Chinese workers had no choice but to comply with the cremation and burning orders. Two hundred laborers gathered the palls and corpses in piles a hundred high. Twenty-two pyres were raised. At 2 p.m. on January 31, 1911, Wu writes, senior medical officials were invited to watch "the first mass cremation of infected bodies in history." When kerosene pumps proved too slow, workers were sent onto the piles with paraffin cans. Soon the site was ablaze; the pyres began to collapse into ground softened by the heat. The next day, bones and ashes were collected and thrown into new pits. Cremation was an evil end, and its value as a preventative measure was far from certain. The Chinese magistrate's office in Hulan was burned during a protest that month by several thousand workers, but the plague fighters did not relent. In February, another fourteen hundred bodies were burned.

Not only bodies but houses and the plague hospital were "consigned to the flames," Wu writes. The only Chinese-speaking Russian doctor in Harbin thought this step unnecessary and cruel, amounting to "destroying other people's property." Steaming, he wrote, was just as effective as burning. (Defoe, too, questioned the use of fire against plague in his *Journal*.) But Wu's methods won out. Much of Fuchiatien and poor areas of Harbin were burned. No funds were provided for rebuilding; they were still poor slums under the People's Republic a half century later.

By springtime the plague had run its course—one in every twenty people in the city was dead—even as hastily disposed corpses continued to be revealed by the melting snow. "Coolies" and beggars ranked first among the epidemic's victims; wealthy foreigners, for all the usual reasons, ranked last. In medieval Europe, plague was called "the poor man's death."

The end of the Harbin epidemic remains something of a mystery. Wu's preventative efforts and the burning of corpses may have limited its spread, but modern historians suspect that some meteorological effect—warming days, changes in humidity—helped to destroy the remnants of the worst outbreak of pneumonic plague ever recorded. It was an outbreak that might conceivably have crossed the globe. Seven years later, the exponentially more contagious (albeit less fatal) "Spanish" flu did just that.

The most definitive of Wu's innovations was probably the design of the mask that became the N95, a pair of which I purchased for the first time in a pharmacy in Kota Kinabalu in February 2020 for about four dollars. For his accomplishments, he received honors from the czar of Russia and the president of France, and in the same year the epidemic ended, 1911, Wu chaired the First International Plague Conference. The conference took place in Mukden and was informally known as the "assembly of ten thousand nations." Its members debated conflicting theories of the disease and its treatment. They were, in the words of one scholar, "effectively deciding which empire was modern enough to rule Manchuria." A few months later, the Qing Dynasty collapsed.

4

"Occidental tourism," the German critic Gerhard Nebel wrote in 1950, "is one of the great nihilistic movements, one of the great western epidemics whose malignant effects barely lag behind the epidemics of the Middle and the Far East, surpassing them instead in silent insidiousness. The swarms of these gigantic bacteria, called tourists, have coated the most distinct substances with a uniformly glistening Thomas-Cook-slime, making it impossible to distinguish Cairo from Honolulu, Taormina from Colombo."

Tourism as epidemic, tourists as gigantic bacteria, coating every surface with the obscuring jelly of Cook & Son, or, in more modern terms, the excretions of easyJet, Airbnb, Opodo, Travelocity, JetBlue, and other corponyms—the comparison is difficult to avoid. Travel and disease are siblings. They contain each other.

I was traveling when the COVID-19 pandemic began, and by the time its gravity was clear, travel had already carried it to every corner of the world. Yet as with the more deadly forms of plague, the affliction Nebel described is not exactly what it seems.

The critique of tourism is as old as tourism, the bacterial metaphor for the tourist almost as old, one could argue, as Antonie van Leeuwenhoek's discovery of the animalcules that wriggled and multiplied beneath the lenses of his Delft workshop. Travelers are like plagues mainly in the obvious sense that both are alive and therefore display the two features common to all living things, movement and metamorphosis. Plagues travel because all of life, in its useless vaporousness and gesticulatory ennui, must seek out novelty. Living things must grow and mutate or else die.

Travel itself has mutated many times over. The invention of European tourism was a mutation of the holy pilgrimage. The word "tourism" first appears, not coincidentally, in English, alongside the rise of the Grand Tour, whose aims were educational as well as diversional. But new kinds of travel do not wipe out former morphological types. Not only does the purposeful nature of travel persist—for migrants, refugees, nomads, sailors, the lovelorn and fugitive, and indeed, religious pilgrims—it is not at all clear that travel was ever merely the product of biological or economic necessity, that there was no pleasure to be had in the novelties of nomadic motion. The human reserve of restlessness was not invented by Thomas Cook.

Even Wu's autobiography is at its core a travelogue. *Plague Fighter* is filled with incident and observation; the reader is treated to a Grand Tour of Manchuria in addition to harrowing scenes of death and lengthy descriptions of medical conferences.

Better to swap Nebel's "travel as plague" for its inverse, "plague as travel": travel from the perspective of virus and disease. What are must-see sights to cilia? What is a tourist attraction but a reservoir—an oasis—of gathered hosts? We might imagine how organic flows of zoonosis are retarded by fresh air and cotton masks just as they are accelerated by coughing and by the pressurized carton of damp bipedal life that is the 747, the 777, or the double-decker Airbus A380.

If no one traveled, there would be few border-crossing epidemics. Zoonotic diseases would be local affairs. Global pandemics like COVID-19 would be impossible to imagine without the flow of human labor and leisure, which are the commercial airliners of communicable disease.

Sex, in particular, is a kind of epidemiological cruise ship. Iberian sailors brought syphilis to Europe. Captain James Cook's crewmen carried measles to the Pacific Islands. Key vectors in the spread of HIV in Africa and India are long-distance truck drivers. As a rule, travelers fuck. A study of 599 Norwegian travelers found that more than 40 percent had enjoyed a sexual encounter while abroad, a majority without condoms. Such are the wages of pleasure-seeking, both for animal life and for whatever equivalent of pleasure drives a bacterium from one moist pit of ingress to another. In 2020, at the height of global quarantines and travel restrictions, a fast-moving quinolone-resistant strain of *Neisseria gonorrhoeae* demonstrated that all national borders are permeable in the dark.

Nations have demonstrated their inability to control the spread of a highly contagious aerosolized disease, but the nation is not the only unit of community on offer. Small bands of men and women living together never lost the cultural memory of plague. They have always known what to do: close the roads, cut off trade, warn away outsiders. Before a global system of capital flow and resource extraction demanded constant access to the most remote hinterlands, a plague was not always the widespread disaster it is today. Villages and tribes could shut themselves away from the local interchange of supplies, forest goods, and spouses. Nomadic communities on the Eurasian steppe could—and did—pick up stakes and flee whenever plague struck. In early 2020, when I visited Indigenous communities in the northern Philippines and the Malaysian rainforests, I found villages open, but soon after I left, as the threat of COVID-19 became apparent, they shut down roads with wooden barricades and warning flags. The Augustinian historian Antonio Mozo observed a near-identical reaction on his travels to Northern Luzon in the eighteenth century. He discovered most people in the mountains unconvertable to Christianity but satisfied to have a handful of siblings or cousins willing to join the faith and act as agents of commerce in the lowlands. These sentinel relatives also helped tribespeople limit their own first-hand contact with the colonizers. "One reason," Mozo wrote, "is the great fear they have of smallpox, which pestilence has never entered their settlements in the mountains." As soon as news of an outbreak reached them, villagers closed all roads and passes with felled trees and underbrush, "and sent out word that if anybody should be so bold as to enter, they will kill him immediately."

The roadblocks had not yet descended that winter when, in Singapore, we had our temperatures taken outside a history museum. Attempts to evade testing were punishable by imprisonment. A cabdriver we met was not taken in by the precautions that governments were putting into place, with new restrictions coming down every few days. "When it's your time, nothing will save you," he shouted. "Not masks, not soap—nothing!" We traveled on to Penang before the first major breakout was discovered in Malaysia, after which a national movement control order grounded us. It was there, in his birthplace, unable to travel and with nothing but time, that I discovered Wu's autobiography.

5

There are still outbreaks of pneumonic plague, particularly around Inner Mongolia where Wu encountered the disease. Sometimes an old man will catch and eat a wild marmot and succumb to fever. If not treated with antibiotics right away, he will die. Some truths are eternal: There will be plague as long as marmots have fleas. When viewed from sufficient geological depths, every landscape is a cemetery.

Sixteen years before the great plague in Manchuria, in 1894, a senior doctor at a hospital in Aksha named Mikhail Eduardo-vich Beliavsky was examining a more limited outbreak of bubonic plague. He came to believe that the disease, which killed dozens of people each year, was being spread by the skinning and butchering of infected marmots. No scientist had ever published a research-based theory of animal transmission of plague. It was four years before Paul-Louis Simond's famous paper on rats. Diagnosis of plague was by no means straightforward in the days before laboratory staining. Plague resembles many other infections.

The notorious buboes are erratic, particularly in pneumonic cases. Beliavsky was mocked.

Among Mongolian hunters on the steppe, however, it was common knowledge for centuries that marmots are a natural reservoir of plague. Hunters could identify subtle signs of illness in their quarry and had even developed a test for plague that involved pricking a killed marmot's paw to see whether the blood had coagulated. If they killed an animal that seemed diseased, they fed it to the dogs, who were thought not to suffer from plague's effects. They knew most deaths took place in winter and that the illness could pass from tarbagan to human through some esoteric medium.

We now know the medium is fleas. In 1910, during the outbreak in Harbin, Wu and his assistants collected two types of blood-sucking arthropods from the tarbagan: a tick of the genus *Rhipicephalus* and a flea, known as *Ceratophyllus silantievi*. This particular tick does not bite human flesh. The flea does. A recently captured tarbagan is usually teeming with both. The arid climate and cold winters of the Mongolian plateau are cosmologically arranged to bring together the marmot, who hibernates in a burrow plugged up with soil, straw, and its own feces, and the flea, whose larvae are born there. Fleas break the skin around every marmot orifice to feed on its blood. When a flea ingests *Y. pestis*, the bacteria multiply until the insect's foregut is overflowing. It then disgorges the bacteria into its bitten host. Inside the marmot's burrow, plague can lie dormant for more than a year.

The plague-carrying marmots of Eurasia do not travel far under their own steam. Fleas, however, are born explorers. Plague-carrying fleas include not only *Ceratophyllus silantievi* but also *Oropsylla hirsuta*. Members of the latter species, infected with plague, were found on the dead bodies of black-tailed prairie dogs for the first time in Denver and West Texas in 1945, and have since been found wherever prairie dogs roam, from New Mexico to Wyoming, usually killing between 80 and 100 percent of an individual colony when they appear. Plague may even be one of the main natural checks on population growth among the social prairie dog, whose colonies, as they grow in size and complexity—the largest ever recorded, in Texas, covered nearly forty times as much land as the city of Houston, and supported a *Sciuridae* population of four hundred million—become increasingly likely to suffer an exotic, vector-borne, epizootic die-off.

As plague travels, so does ritual. Traditional hunters are attuned to sentinel marmots, those lookouts who stand erect at the portal of the burrow and warn the colony inside of a predator's approach. Their behavior is diagnostic. Sluggish or unsteady marmots suggest that the whole colony might be diseased and should be avoided. According to oral tradition, the sentinel marmot of a sickened colony will set out on a journey to find medicinal plants. Such marmots, observed on their own, far from any burrow, are likewise not to be hunted.

The day we first entered Malaysia, February 2, is celebrated in the United States as Groundhog Day. The holiday came to the New World with the Pennsylvania Dutch, who imported it from Germany, where it is Christianized as Candlemas and has long blended with the ancient Celtic holiday Imbolc. There are passage tombs in Ireland more than four thousand years old whose mouths point to sunrise around Imbolc Day.

Although I have never come across a direct connection, it has struck me that Groundhog Day might bear some obscure or even shamanistic connection to plague. Could the divination of spring be related to the hoped-for dissipation of the classic season of disease? Is the groundhog of American tradition related to the sentinels of the steppe? In ancient Imbolc, later St. Brigid's Day in Ireland, the divining creature was often a badger or a serpent, whose burrow or hole would be watched for signs of life, as in the Scottish Gaelic proverb:

Thig an nathair as an toll
Là donn Brìde,
Ged robh trì troighean dhen t-sneachd
Air leac an làir.

The serpent will come from the hole
On the brown Day of Brìde,
Though there should be three feet of snow
On the flat surface of the ground.

Candlemas, St. Brigid's Day, Imbolc: all holidays grouped together at the midpoint between winter solstice and spring equinox, an unstable seasonal fulcrum. Another couplet makes clear the eternal return whose awesome possibility lies within the day's powers of divination: "If Candlemas Day is bright and clear / There'll be twa winters in the year."

There is even a weather-divining, disease-preventing Christian saint named Quirinus, a Roman tribune from Neuss whose name has been invoked against smallpox, gout, and, indeed, bubonic plague. In addition to his protective powers against disease and pandemics, Quirinus is a patron saint of animals. His feast day, March 30, is likewise a day of divination for German farmers, some of whom still recite the adage *Wie der Quirin, so der Sommer.* As goes St. Quirinus's Day, so goes the summer.

These scraps of folk belief combined with others in the dim past to form the contemporary American holiday. Today we raise the groundhog—the species *Marmota monax*—and hope that this sentinel will not condemn us to more winter; in other words, that the meteorological conditions which long sheltered *Y. pestis* in Asia will dissolve into the wet life of spring.

Weather-divining animals have been found in traditions across Europe since long before the outbreak of plague witnessed by a five-year-old Defoe. For Candlemas, in Germany, the role has been filled by badgers, foxes, and even bears. ("If the bear can see over the mountain," one description reads, "he must spend another eight weeks in the hole.") As described in the poetry of the Tang Dynasty, the behavior of pigs was once thought to predict rain. And in tribal communities all over the world are found feast days and rituals intended to ward off the random and cataclysmic illnesses that would occasionally decimate a population. It is thought the Mongol Empire brought plague to Europe, as Chinggis Khan's son Ögedei rode his conquering armies across the regions now called Ukraine, Bulgaria, and Hungary. The nomadic empire is theorized to have spread *Y. pestis* through trade caravans carrying marmot pelts, leathers, and bags of millet (with possible rodent stowaways). Mongols also carried shamanistic traditions across the steppe and mountains to the arctic edge of Europe. In several Mongol legends, the marmot originated as a man who, failing to shoot down the sun, cut off his thumbs and became animal. He remains a liminal figure between house and wilderness, between animal and man. The marmot and its shadow. The conjunction is nicely captured in *quarry* itself, a word that by accident of etymology means both prey and excavatory source.

Under the orderly signs of scientific enlightenment, which indulge only playful superstitions, we allow the groundhog to "predict" the continuation of winter when he is lofted like the Sabbath cup before the men and women of Punxsutawney. But it takes only a small movement of faith to imagine him as a fleshy vessel for the spirit-god who was spring's conjurer and death's defeat. There are tokens hidden in his shadow. It may be pure speculation—a tourism of the mind—but my thoughts go to plague, illness, and miracles each February, as the raising of the marmot flattens the crowd to silence.

ISOBEL WHITCOMB

An Ark for Amphibians

FROM *Sierra*

THE FOOTHILL YELLOW-LEGGED frog is a slippery creature about the size of a silver dollar, with topaz eyes and neon-yellow splotches on the underside of its meaty legs. During the approximately six to nine million years of its existence, the amphibian has stood sentinel to change in the Sierra Nevada. It watched as glaciers encrusted its habitat in ice and then retreated. When European settlers blasted away riverbeds in search of gold, the frog hung on deep in the rocks and crevices of tiny streams. It persisted even as dams corralled its waterways.

"You list all these things, and they're still here," said Ryan Peek, an ecologist at the University of California, Davis. "They are amazingly resilient."

Global warming could change that for good.

On a scorching-hot morning in mid-August 2021, Peek and I crouched in the steep, oak-shaded Robbers Ravine. Air tankers circled overhead as they prepared to drop fire retardant on the nearby River Fire. Nestled in the gully were a series of deep tubs carved into the bedrock over time by rushing water. The creek through the ravine traces a path to the North Fork of the American River, which in turn rushes to the Folsom Reservoir and on to Sacramento. That day, the pool into which we stared was empty. Other pools contained shallow, murky water or slurries of green algae. "This is crazy," Peek said.

Fifty years ago, foothill yellow-legged frogs crowded the banks of streams throughout much of California. Scientists describe how it used to be difficult to take a single step without squishing a frog beneath your foot. "They were just like popcorn," Peek said. But in the second half of the twentieth century, the frogs began to disappear. Non-native bullfrogs invaded their habitats and ate them. Chytrid fungus—the source of chytridiomycosis, a deadly disease implicated in the extinction of ninety amphibian species—swept through some watersheds in Southern California, killing off the frogs. Then, in late 2011, California entered its most severe drought in as many as twelve hundred years. Tributaries like Robbers Ravine dried up, crowding the frogs into shallow pools. This created the perfect conditions for the fungus to spread. The frogs became an easy feast for predators.

Foothill yellow-legged frogs have disappeared from half their former range, and the population is declining fast. "Knowing that some of these places may not have frogs when my kids are adults sucks," Peek said. "It's not a great outlook." In 2019, thanks to Peek's graduate dissertation on the species, five of the six subpopulations of foothill yellow-legged frogs in California were listed as endangered.

Heading toward the top of the ravine, we reached an oasis—an unexpectedly lush pool fed by groundwater. "This is my spot," Peek said. Despite the apparent lack of water in Robbers Ravine, the place was peppered with frogs. They leaped off the steep walls of the gully. Beady eyes peeked up at us from the remaining puddles.

It's the prospect of oases that gives Peek and other scientists hope, for not just the foothill yellow-legged frog but also myriad other species, from New England's red squirrels to Oregon's elk. Climate change doesn't spread its effects evenly over the landscape. While some locales are pummeled by extreme weather year after year, others remain curiously unaffected. These places remain cooler than the surrounding landscape, and quirks of geology buffer them from the effects of global warming. Scientists call these safe havens *climate refugia*. They hope these pockets in the landscape will provide plants and animals an ark in which to ride out the worst effects of climate change, until conditions become favorable once more.

But first, scientists have to find the refugia and protect them—and get the plants and animals there in a process of assisted migration. A coalition of scientists, conservationists, and land managers in California are beginning that process for many species, including the pika, the Torrey pine, and a certain frog with splotches of neon yellow on its legs.

Long before humans heated up the atmosphere with greenhouse gases, shifts in climate shaped the course of life on Earth. Over the past 2.6 million years, roughly 30 percent of Earth's surface became encrusted in ice, then thawed, then froze again—around 20 times. Just about all species on Earth were shaped by these cycles. Many wouldn't exist if their ancestors hadn't found refugia.

The term *climate refugia* emerged from paleoecology, the study of how plants and animals interact with their environment and one another over time. For hundreds of years, scientists have debated the reason for the incredible diversity of species on Earth. Why, for example, do we see eighteen species of macaw in the Amazon rainforest instead of just one? While Darwin's theory of evolution supplied a broad intellectual framework for speciation, it left many of the details unanswered. In the 1960s, some researchers proposed refugia as a possible answer. They hypothesized that during ice ages, species had escaped to pockets that remained hospitable. In refugia, plants and animals continued to thrive until the climate warmed once more and they were able to fan out across the landscape again. These periods of isolation allowed plant and animal populations to evolve separately, until they were no longer the same species. The role of refugia in these species' survival is written into their genes.

Around the turn of the millennium, *climate refugia* began to be used in the context of climate change wrought by humans. As scientists began warning the public about global warming, some also started to wonder where plants and animals might have the best shot at survival. But they had no way of precisely projecting into the future, said Carlos Carroll, an ecologist at the Klamath Center for Conservation Research. In the early 2000s, researchers got a new chest of tools: computer programs that could analyze myriad variables about the environment, from topography to soil composition, and better calculate the future of the climate in a particular region in more granular detail.

These models presented scientists with a new understanding: "The stress from climate change, the level of risk to species and ecosystems, is not uniform across regions; it varies quite a bit," Carroll said. "That was the key finding."

But what gets classified as climate refugia? Bumblebees in Europe and chipmunks in Yosemite are seeking cooler climates at higher elevations. For some scientists, that alone is enough to qualify a mountaintop as a climate refuge. Others disagree. After all, where will those plants and animals go when even those mountaintops become too warm? For Toni Lyn Morelli, a research ecologist with the US Geological Survey's Northeast Climate Adaptation Science Center, refugia are places where the climate will change very slowly, or won't change at all.

Take Devils Postpile National Monument, in a valley in the central Sierra Nevada. Even in the heat of summer, the meadows in the valley remain chilly—as much as 18 degrees cooler than the 10,000-foot peaks around them. A phenomenon called cold-air pooling provides a buffer against heat waves. During the day, as heat rises, cold, moist air settles to the valley bottom. Meanwhile, the steep walls of the monument prevent sunlight from warming that air, so the coolness gets locked in. Across the globe, other refugia form around steep slopes that face the poles—the lack of sunlight means that these areas warm less readily. Deep lakes and coastlines form their own refugia, with some areas absorbing heat while others stay cool.

The harsh reality is that it might not make sense to invest limited resources in protecting environments where species will no longer be able to survive. "There's much, much more work to do than we have resources," Morelli said. That work includes removing invasive species and reducing humans' impact. It might even include moving vulnerable species to refugia.

Assisted migration is not without controversy. Some land managers are concerned that newly introduced species will edge out native species, or that newly established populations will eventually blink out on land they haven't evolved to inhabit. Alexis Mychajliw, a paleoecologist at Middlebury College, points out that plants and animals aren't stationary beings, and during past climate change events, they moved themselves. "The history of most species is movement," Mychajliw said. "When the climate changes, they tend to follow those conditions." The sheer pace of human-caused climate change, plus the barriers posed by roads, fences, and subdivisions, make that natural pattern of movement impossible. So why not give species a hand if we know where they might have a better shot at survival?

In 2017, ecologists at the National Park Service introduced a cohort of threatened red-legged frogs into Yosemite after identifying areas of the park as likely refugia. The species once lived throughout California but is now limited to pockets across the state. Whether the site of the frogs' introduction was ever a part of their original range is contested, said Andrea Adams, a research scientist at Washington State University, but today, the red-legged frogs are thriving in Yosemite. Adams considers it a conservation success.

It's possible that a similar strategy might save the foothill yellow-legged frog, but first, land managers need to know where the climate refugia are. That's where Claudia Mengelt comes in.

Mengelt, an oceanographer turned land manager, had never heard the term refugia until she stumbled upon Morelli's research while working at the National Academy of Sciences in Washington, DC. Mengelt, who is now the program manager of land management research at the US Geological Survey, was immediately intrigued. Because of the fluid, dynamic nature of ocean environments, marine conservation tends to embrace the inevitability of change and movement, so the idea of climate adaptation felt natural to Mengelt—a fish out of water among frog biologists. By the time she met Morelli in person, Mengelt had moved out West and taken on a new project: developing a conservation plan for the foothill yellow-legged frog.

"I'm like, let's get the two projects together," Mengelt said.

The deeply rutted dirt road into the Tuolumne River canyon is a series of hairpin turns through sage and dusty shrubs. "This is one of those incomprehensible roads the Forest Service built way back in the day," said Steve Holdeman, a forest aquatic biologist for Stanislaus National Forest. He was driving Mengelt and me down into the canyon to check out some of the sites where Mengelt's team is doing research.

At the bottom of the canyon, we stepped out of the car into blistering heat and air that smelled like campfire smoke. Burned trees lined the hillsides above us. Below, the South Fork of the Tuolumne River wound its way leisurely around rocks and boulders. Just like the pools at Robbers Ravine, the river was thick with algae and white scum. "River snot," Holdeman said. The recently low streamflows weren't strong enough to wash the stuff away. "We're down a couple of inches every year," he added.

While Holdeman isn't directly involved with the refugia mapping initiative, the project's findings could affect the management of Stanislaus National Forest and in turn the work he does with foothill yellow-legged frogs. "Claudia and all those bigger-brained people—they have a more strategic approach," he said in his Tennessee drawl. "I've got the bigger picture on these lands."

A central goal of Mengelt's work is to overcome what she calls the "knowledge to action" barrier. The traditional model of ecology—testing hypotheses, developing conclusions, writing up findings, then going back to the drawing board—is a lengthy process and often fails to present a clear set of actions to policymakers. With climate change, we don't have that kind of time.

"Sometimes the information just describes the problem," Mengelt said. She wants to present land managers at the National Park Service, the Bureau of Land Management, and the US Forest Service with science-based solutions. Together, they could develop decisions based on the data this project generates, even as it evolves.

The project is focusing on two river systems: the Tuolumne and the Merced. Both cascade down from the High Sierra through Yosemite before winding their way into the public lands of the Sierra foothills. The work happening at this particular site on the Tuolumne is part of the first prong of Mengelt's project: determining where the frogs are.

That effort is led by Adams. Like Peek, Adams is recording population counts of the frogs she sees, but she's also using a technique that wouldn't have been possible fifteen years ago: environmental DNA analysis. Adams is testing water samples for traces of the genetic material left behind by foothill yellow-legged frogs. Meanwhile, a separate team of scientists is working on developing a computer model to identify potential refugia. Tina Mozelewski, a postdoctoral researcher in Morelli's lab at the University of Massachusetts Amherst, is leading that initiative, which involves layering environmental data—such as streamflow and vegetation—on a map alongside population data for the foothill yellow-legged frog. The goal is to determine which variables best predict where frogs live and might be most important for their survival. The scientists will combine the habitat data with previous calculations on the rate of climate change across the landscape, which take into account variables like cold-air pooling. By crunching this data, they can estimate where conditions will change the least over the next one hundred years in the Tuolumne and Merced watersheds.

For foothill yellow-legged frogs, refugia will likely be in places where streamflow continues steadily despite drought, said Sarah Yarnell, a hydrologist and river ecologist at UC Davis. They'll likely survive in spring-fed pools—like the one we saw at the top of Robbers Ravine. Springs are fed by aquifers that build up over time as snowmelt and rain leach through cracks in the bedrock. For the time being, the flow from springs is buffered from fluctuations in rainfall or snowmelt. Plus, this water is colder: "Those locations are going to be the most climate buffered because they're not going to be exposed to warming air temperatures," Yarnell said.

We talked about refugia in aquatic environments, and the conversation naturally turned to dams. While the lackadaisical South Fork of the Tuolumne and North Fork of the American are both unregulated by dams, the main stem of the Tuolumne is fragmented by them. It descends from the O'Shaughnessy Dam, which creates the Hetch Hetchy Reservoir higher up in the Sierra, and rushes toward the Don Pedro Dam in the foothills. Every day between 7 a.m. and 11 a.m., O'Shaughnessy Dam releases a pulse of water so thrill-seekers can raft down the turbulent flow. This unnatural rhythm in streamflow isn't ideal for foothill yellow-legged frogs. They're highly attuned to changes in hydrology—the pulse of snowmelt in late spring is their cue to breed. Pulses of water wash egg masses downstream. Plus, the frogs have a hard time migrating around the giant reservoirs created by dams. Isolated in pockets by the dams and unable to live in the cold, deep reservoirs, they've lost the genetic diversity to keep their populations healthy.

How these dams will interact with climate change is a major question. Peek suspects that river sites without dams will function as a kind of climate refugia in part because they will allow frogs the mobility to seek more favorable climate conditions on their own and the genetic diversity to withstand changes in their environment. Other researchers wonder if dammed rivers could function as an unnatural climate refugia—dams release water from the bottom of reservoirs, where temperatures remain cold due to lack of sunlight. Plus, the streamflow beneath dams will be somewhat buffered from drought.

Both arms of the foothill yellow-legged frog project—the population data gathering and the refugia mapping—are useful only in tandem. Paired with population data, refugia mapping helps researchers understand which sites will support future conservation efforts, including introductions, and which won't. "When you're talking about reintroductions, that has to be part of the conversation," Adams said. Frog populations that are large and healthy now might serve as sources for future translocation projects—scientists might collect eggs or tadpoles and rear them in a zoo before releasing them at another site. And places that look like refugia might be important sites for those introductions, whether or not foothill yellow-legged frogs ever lived there. "If the goal is to get frogs out on the landscape and save the species, it doesn't really matter," Peek said.

It's possible that if the climate changes enough, even refugia will reach a sudden tipping point when their regular buffers will no longer function. If, over hundreds of years, snowmelt no longer replenishes underground aquifers, they will eventually empty. Cold-air pooling might not happen in hotter conditions. No refugia will last forever, Morelli said. "I would just say, we don't know if there are tipping points."

During past climate change events, species had more time to evolve traits that would allow them to weather the new conditions. But human-caused climate change is happening at an unprecedented rate—too fast for many species to adapt. Even having just a thousand extra years would make a big difference. "We need to protect the places where things are moving at a slower rate so that species can do their thing as they have for millions of years," Morelli said. And we might not even need that much time. A recent report from the Intergovernmental Panel on Climate Change found that if we were to significantly reduce carbon emissions, the climate would stabilize within decades.

Not far downstream from where we parked in the Tuolumne River canyon, the South Fork meets the Middle Fork in a deep pool. Holdeman waded in thigh-deep and dunked his head. Mengelt let herself fall backward, fully clothed, into the current so that only her head and feet stuck out. The cold water was a welcome relief from triple-digit temperatures.

Soaking wet, we picked our way back up the South Fork to the car, then trundled up the canyon road. As we climbed the switch-backs, we passed a stream tucked into a fold in the hillside. Holdeman pointed out that all the surface water had long dried up—this stream was likely fed by groundwater. "These little tiny things are incredibly important refuge habitats for the frogs," Holdeman said.

He was reminded of the time when he and Adams had climbed to a waterfall that cascaded down a sheer rock face. At the top, they were confronted by an unlikely sight: a foothill yellow-legged frog. "A beautiful golden one," Holdeman said. It had climbed up the thirty-foot waterfall in search of better habitat. Here, among the beige shrubbery, the stream stood out, kelly green, lush with ferns and moss—an oasis.

LOIS PARSHLEY

Don't Look Down

FROM *Grist*

KATHY LENNIGER WAS running her dogsled team one day along her usual route in Fairbanks, Alaska, when she suddenly splashed into overflow, fresh water spilling on top of the snow. Surprised and chilled, she returned to the parking lot, where a lanky man was loading a sled with science equipment. Nicholas Hasson, it turned out, was studying thawing permafrost—research that could shed light on the streams and sinkholes that recently materialized around Lenniger's property and all around town.

Lenniger lives in a log cabin in Goldstream Valley, a spruce-lined swale with a rolling view of the Northern Lights near Fairbanks. "It's the birthplace of American permafrost research, actually," said Hasson, a PhD student at the nearby University of Alaska Fairbanks, or UAF. During World War II, the military feared the ribbons of dancing light were interfering with its radar, so Congress passed an act in 1946 establishing the Geophysical Institute at UAF. Soon scientists were investigating the strange phenomena in the sky and drilling boreholes around Goldstream Valley to study the frozen ground beneath their feet.

Since then, temperatures in Fairbanks have shifted so much that the National Oceanic and Atmospheric Administration officially changed the city's subarctic designation in 2021, downgrading it to "warm summer continental." As the climate warms, the ancient ice that used to cover an estimated 85 percent of Alaska is thawing. As it streams away, there are places where the ground is now collapsing. Many of the valley's spruce trees lean drunkenly. Sometimes only a thin layer of soil covers yawning craters where the ice has vanished, what Hasson calls "ghost ice wedges." Its absence has already fundamentally changed how—and where—people can live.

When Lenniger built her cabin several decades ago, she didn't expect she'd need to regularly jack up her foundation. But for the last several years, she said, "if I have some water on my counter now, it rolls in this direction. It's like, 'Oh yeah, it's sinking again.'" At first, she tried to fill the sinkholes popping up around her property with bones from the meat she fed her sled dogs, but eventually the pits grew large enough to strand a backhoe. Despite living in perhaps the most-studied permafrost valley in the country, Lenniger didn't know how much worse her troubles might get—until Hasson offered to help.

On a muggy afternoon last summer, Hasson prepared to try to find out why Lenniger's cabin was sinking. He pulled on a backpacking frame he'd jury-rigged to receive very low-frequency radio waves from antennae in Hawaii, recording the modulations of the electric field to map the permafrost beneath the duff. The colors of the aurora come from the charged particles of solar wind, which collide with oxygen and nitrogen in the Earth's ionosphere and create a glowing halo. The free electrons from these collisions can reflect radio waves, helping Hasson understand how permafrost is thawing below the surface. Combined with a $40,000 laser he dragged behind him on a plastic sled he'd nicknamed "The Coffin," Hasson is able to link surface methane emissions to the ice disappearing underground.

As he scrambled off Lenniger's driveway into the brush, Hasson explained, "It's just like an MRI—we're able to scan and see where water is flowing." Walking across her yard, he found a new underground river had formed under a corner of Lenniger's home, which explained why her land had caved in.

The permafrost around Fairbanks is discontinuous; jagged pieces of it finger north-facing slopes and enfold the low-lying valleys. Yet potential home buyers who want to avoid it are left to guesswork. "There's no comprehensive map of permafrost," said Kellen Spillman, the director of the department of community planning for the Fairbanks North Star Borough. For those like Lenniger, whose properties later develop thaw-related problems, there's little recourse, either from insurance or the government. The University of Alaska Fairbanks, home to much of the state's permafrost research, has itself struggled with recurring sinkholes on its roads and parking lots. "We have invested funding to rebuild," said Cameron Wohlford, director of design and construction at the school's facilities, "only to have them fail."

Homeowners around Alaska's second-largest city are facing expensive repairs, or even having their properties condemned. Hasson eventually traced the river running beneath Lenniger's property to her neighbor's, where the owner, Judy Gottschalk, reported that her septic pipes had broken as the ground settled. "My well went out this winter, too," she said. Not knowing where else the ghost ice lies, Gottschalk has been nervous about putting in a new septic system. The drilling and construction required to replace it would cost her as much as $45,000, more than she originally paid for her house. "Everyone I know is having problems with their housing," Lenniger said.

As parts of Alaska set record high temperatures in December, Fairbanks closed out 2021 with a destructive ice storm, causing roofs to collapse. A warmer Arctic is also a wetter Arctic, accelerating the breakdown of permafrost, explained Tom Douglas, a senior scientist for the US Army Corps of Engineers' Cold Regions Research and Engineering Laboratory, in Fairbanks. "For every centimeter of rain, we see about one centimeter of additional top-down thaw," he said. On average, Fairbanks now sees about five more weeks of rain than it did in the 1970s.

"In my forty-seven years here, I've never seen these kinds of conditions before," Lenniger said. She has a lot of practice finding creative ways to take on Alaska's hurdles: Before phone lines went in, she and her partner used homing pigeons to communicate while mushing, though she said she was unfazed when the birds were devoured by owls. But now, the rapid changes are testing her ability to cope. "Every day, it's like, now what will happen?"

Just as the earth clings to its former shape, leaving a record of where ice used to be, the very language used to describe these changes is revealing. The word *permafrost*, after all, is simply an abbreviation of "permanently frozen ground." Much of Alaska's permafrost is tens to hundreds of thousands of years old, first frozen when Goldstream Valley was grazed by mammoths. Now that sense of immutability is slipping. "It was thought to be *permanent*—that any changes happened on a scale of tens of thousands of years," said Vladimir Romanovsky, a professor emeritus of geophysics at the University of Alaska Fairbanks and a leading permafrost researcher.

Many variables influence permafrost's stability, like how cold it is, how deep it runs, and the quantity of soil moisture, or its "ice richness." In some parts of Alaska, ice extends nearly a half-mile below the surface, while in others, it has formed the landscape itself, sprouting tundra-covered ice hills called pingos.

Since 1993, Romanovsky has been taking field data from stations around the state, recording their increasing temperatures. At all of the 350 stations, soil temperatures have warmed substantially, and thaw is inching down to deeper depths. On the North Slope, one of Alaska's coldest ice-rich regions, "when we started, it was about –8, and now it's –4 degrees Celsius, so we're already halfway to zero," he said. Dramatic changes will increase once this melting point between frozen and liquid is hit. He predicts that within forty years, the Slope will be "at a critical threshold in normal, undisturbed conditions."

Off the North Slope, this tipping point will be reached sooner. Any time soil or vegetation is disturbed—as the Army Corps of Engineers discovered in 1942 while trying to build a highway to Alaska—permafrost has a tendency to disintegrate into truck-swallowing mud. It's a similar story with roads built in recent decades. Jeff Currey, materials engineer for the northern region of Alaska's Department of Transportation & Public Facilities, explains that as ice wedges degrade under the state's highways and airports, the asphalt heaves and drops, creating a dangerous roller-coaster effect. Because Alaska has relatively few roads across its 665,000 square miles, the ones it has are critical connections.

"Warming temperatures are contributing to increasing maintenance and damage," Currey said. "Anecdotally, we're having to fix the same places more frequently, and more intensively."

Mitigation measures can help, from the low-tech approach of using gravel to channel cold air against embankments to high-tech thermosiphons, tubes that channel warmth aboveground during the winter to help keep the soil frozen. But Alaska's budget for maintenance is largely dictated by the state legislature, and Currey calls the annual $330 million allotted to the northern region in recent years "inadequate." Currey explains the average road is typically built to last around thirty years, but that's largely based on expected traffic, not whether the road will be thermally stable. An independent study published in the *Proceedings of the National Academy of Sciences* estimated that, as a result of climate change, the state will have to spend billions more on maintaining and repairing public infrastructure by the end of the century. Despite budget shortfalls, Currey predicts that "maintenance efforts simply have to increase." In many cases, "we'll tolerate rougher and worse roads than we do now—that will just be the economic reality."

Around Fairbanks, elevating buildings to keep their heat from leaking into permafrost or designing structures to be adjusted isn't new. Releveling houses as a cheap way to adjust to moving ground is an Alaskan tradition. "My grandparents used to chase the corners on their cabin when it moved, like everybody," said Aaron Cooke, an architect and researcher at the National Renewable Energy Laboratory's Cold Climate Housing Research Center, who has worked on these issues in many communities around the state. But with climate change, the old engineering tricks that helped keep permafrost frozen aren't sufficient. "The ground is changing, even if you do everything right."

To understand the scale of the impact when it starts to melt, Cooke said, you have to understand that "to someone in the north, the natural state of the ground, the default status of Earth, is frozen. And thousands of years of culture are built on that knowledge." While the impacts of permafrost thaw—subsidence, flooding, sinkholes, and landslides—mimic the devastation of natural disasters, the Federal Emergency Management Agency isn't responsible for permafrost damage, and it's difficult to get covered by homeowners' insurance. "How fast does a disaster need to move for a department that handles disasters to address it?" Cooke asked.

Romanovsky predicts that within a decade, the destruction in most parts of Alaska will get worse. "I'm worrying about my house as well," he said. But regions with continuous and ice-rich permafrost, like those in northwest Alaska, will see the worst damage. "It will be the major problem driving relocation," he said, "but these changes need to be understood at high resolution—for each village, for each house, you need to know what to expect."

Where the Chukchi Sea bites into the North American continent, ice loss has driven thousands of walruses to the beaches of Point

Lay, in northwest Alaska within the Arctic Circle. The predominately Iñupiat community, home to around three hundred people, is wrestling with the loss of ice, too: In 2016, the lake they relied on for drinking water disappeared overnight after the ice wedge it rested on eroded, forcing the town to pump water from a nearby river. This year, one of the town's holding tanks failed, spilling almost a million gallons. "Apparently, permafrost was melting under us," said Lupita Henry, the Native Village of Point Lay's former tribal president. "There are cracks in homes, doors that can't close, houses that are so angled they seem unlivable."

Now forty years old, Henry was a young girl when the town's first underground sewer lines were put in; many of them have since broken as the ground settled. The borough government recently installed new electric poles, which are already starting to lean. Like in many rural Alaskan communities, there's a shortage of housing, but Henry said the thawing permafrost makes it difficult to build or even get a loan for a new home. "Where do you get your insurance? Through which bank can you finance to even get your home fixed?" she asked. "When the ground is falling underneath you, what do you do?"

In 2018, the state recognized a new hazard: *usteq*, a word from the Alaska Native Yup'ik language that describes the catastrophic land collapse stemming from thawing permafrost, and the erosion and flooding it entails. As sea ice disappears, the coast has been battered by intensifying storm surges, speeding the breakdown of permafrost under the shore. Riverbanks are corroding from thaw, changing everything from the chemistry of the groundwater to its distribution and movement. Permafrost, Henry said, "is linked to everything—our homes, water sources, food sources, vegetation."

Point Lay is now working with researchers on a Navigating the New Arctic project, funded by the National Science Foundation, to try to determine the best engineering for building on its ice-rich and unstable ground. It's all complicated by the fact that the remote town can be reached only by plane or barge, making construction more difficult. Even before the pandemic, supplies were regularly delayed. "All of the problems overlap," said Jana Peirce, the project's coordinator. Point Lay can apply to FEMA's Building Resilient Infrastructure and Communities program for help in adapting to permafrost thaw; the federal agency is now proactively trying to intervene, because the cost of responding to emergencies is, on average, six times more expensive than mitigation. But to do so, Point Lay will need an up-to-date hazard mitigation plan, and to form a plan, they need to know where the ground ice is, and how it might melt. "While there is no question that planning is important for smart adaptation," Peirce said, "for a small community already living in crisis, this is just another hurdle."

In the 2019 Alaska Statewide Threat Assessment, which set out to summarize the risks permafrost presents, Point Lay is ranked as one of the top three communities under threat from permafrost thaw. Yet aid has been slow in coming. "You tell them you need a water source, that your land is melting underneath you—how many meetings do I have to have until I'm given funding?" Henry asked. In March 2022, Point Lay became the first town in Alaska to declare a climate emergency, acknowledging the threat to their existence.

Towns across Alaska are facing similar challenges: The statewide threat assessment found that 89 of Alaska's 336 communities are threatened by permafrost degradation. "The main barriers to addressing these threats include the lack of site-specific data to inform the development of solutions, and the lack of funding to implement repairs and proactive solutions," Max Neale, senior program manager for the Alaska Native Tribal Health Consortium, wrote in an email. "We have yet to see significant engagement from state and federal partners to improve the efficacy and equitability of programs for communities facing climate change and environmental threats."

In 2020, the US Government Accountability Office found that federal assistance for climate migration has been ad hoc, and that the federal government is nowhere near prepared for the scale of relocation required. Cooke says temperate parts of the world simply don't seem to have registered the urgency of Arctic change. He's spent over a decade racing to help relocate Alaskan towns like Shishmaref or Kivalina, which, despite being deemed in imminent danger back in 2003, have not yet completed their move. But when he attends climate change conferences, "it's very jarring to hear people still talking in the future tense."

For many Alaskans, the emergency is already here. "If we can get a good idea of how much permafrost we're sitting on top of," Henry said, slowly, "we can try to get the federal government to help us with mitigation, or decide if we have to relocate." Although facing a crisis, people in Point Lay are used to the idea of building for an uncertain future. "Don't put any pity on us," Henry said. "We're strong people who survived thousands of years—and we will continue surviving."

The scale of the problem is daunting, but there's surprisingly little agreement on how much dealing with a thawing Arctic will cost. Over a dozen experts interviewed for this article admitted they weren't sure how many Americans live on permafrost; a recent paper published in *Population and Environment* suggested a ballpark of around a hundred and seventy thousand people. Nor can anyone agree how much ice is where, much less how it might thaw.

Nearly a third of Arctic research is based on data from just two field stations: Abisko, Sweden, and Toolik Lake, Alaska. And researchers usually collect data during the Arctic's short summer field season, even though winter conditions may look very different, making conclusions less accurate. For instance, recent studies have found that emissions of carbon and methane released by thawing permafrost have been drastically underestimated. There are 1.6 trillion metric tons of carbon currently stored in permafrost—twice what's now in the atmosphere. New projections suggest that the amount of greenhouse gas emissions from permafrost could equal those emitted from the rest of the United States by the end of the century.

"It's clear that the models are not capturing all the key pieces," said Anna Liljedahl, a climate scientist at the Woodwell Climate Research Center, who is based in Homer, Alaska.

Research attempting to settle these questions generally falls into one of two camps. There's top-down, like Liljedahl's work with the Permafrost Discovery Gateway, which uses high-resolution satellite imagery to record thaw slumps and surface water changes. Machine learning and supercomputers have helped Liljedahl closely map visible ice wedges, creating a more comprehensive view of the Arctic, but can only infer what's under the surface through identifying types of soil or vegetation.

The second approach is bottom-up: Romanovsky's boreholes, for instance, deliver very detailed measurements from specific places, but researchers have to extrapolate to draw larger conclusions. Yet all permafrost is not equal. Take a type of permafrost called yedoma: frozen, silty muck from the Pleistocene era that releases ten times more greenhouse gases than other types of thawing ice. Additionally, research indicates yedoma-rich regions may be warming the most quickly. So knowing how much yedoma there is, and where, is critical.

Scientists like Hasson hope to advance a third approach using airborne imaging spectroscopy, essentially mounting a fancier version of the laser on his sled to planes, a more efficient research method. This technique can detect large methane emissions, and Hasson can then use very low radio frequencies to identify what's happening below the surface, identifying methane hot spots and providing information on the scale that infrastructure planning requires.

"The question is, why aren't we doing this method at scale?" Hasson said. "Why am I not in a plane right now flying over Alaska?" The Department of Energy considers permafrost thaw and its emissions a threat to national security, and is partly funding Hasson's research, along with NASA and the National Science Foundation.

Much is at stake. Dmitry Streletskiy, a geographer at George Washington University, explained that long before ice begins to thaw, warming decreases permafrost's ability to support structures. In the spring of 2020, the 800-mile Trans-Alaskan Pipeline reported its first instance of "slope creep," as thawing permafrost jeopardized its structural integrity. That's likely what happened in the Siberian city of Norilsk a few months later, where thawing ruptured a huge fuel reservoir, prompting a cataclysmic diesel spill that dyed the region's rivers blood-red.

Streletskiy started his career focused on ecosystems, but realized that "unless you put monetary values to things, it doesn't get much attention." His most recent study found that 70 percent of major Arctic infrastructure is in areas that permafrost thaw could put at risk of damage within the next thirty years, increasing maintenance costs by $15.5 billion, as well as causing another $21.6 billion in damages. And those are the paper's most conservative estimates.

While Russia likely has the lion's share of the world's population living on permafrost, alpine countries like France and Switzerland will also see mountain slopes start to lose their stability, resulting in hazardous landslides. A recent study published in *Population and Environment* found that 3.3 million people currently live in settlements where permafrost will degrade by 2050, forcing many to relocate.

"Those who live on permafrost have a pretty good understanding of what will happen in twenty years—they don't need scientists to tell them," Streletskiy said. "It's the people who live in DC or Moscow who need to pay attention."

Up the rippling highway from Lenniger's cabin in Goldstream Valley, Sam Skidmore shoveled dirt away from a vault door at his gold mine, the entrance to the deepest permafrost tunnel in Alaska. He'd decided to break his rule against opening it when the temperature was above freezing so Hasson could take ice samples. Skidmore stumped down into the darkness, his headlamp gleaming off ice crystals as he passed a woolly mammoth skull poking out of the wall. As they continued deeper, gravel beds betrayed warning signs of past eras, when dramatic warming transformed the landscape. "We're literally walking back in time," Skidmore said.

They descended between alternating layers of gravel and silt, passing eons when interior Alaska was an endless grassland steppe and eras when a changing climate shaped the landscape into more familiar forests. "Where we are now [in time], *Homo sapiens* hadn't entered America," said Skidmore, who is preserving the tunnel for research. He poked at a particularly pebbled section, saying it would take "a horrendous amount of rainfall to take all the trees and silt away and make a new layer of gravel like this."

Today, the Arctic is again confronting dramatic change: As the region's permafrost continues to thaw, some areas of Alaska will sink and get wetter, while others may dry out and burn, transforming habitats. Other studies show that permafrost under the ocean itself is thawing, reshaping the seafloor, forming craters the size of city blocks and elevating new pingos. For humans and animals alike, responding will be a balancing act, said Dmitry Nicolsky, a research associate professor at the Geophysical Institute in Fairbanks. Hazards will combine to create cumulative effects: As wildfires increase, for instance, people are told to cut vegetation away from their houses. "But making a safety buffer in Fairbanks might also cause permafrost degradation," Nicolsky said.

Almost above Skidmore and Hasson's heads, on the other side of the tunnel's glistening roof, was one of the countless lakes that dot Alaska's interior. In January at 40 degrees below zero, Hasson can drill into its frozen surface and light the escaping methane plumes into towering columns of fire. The lake is also releasing mercury, a toxic metal that could now be accumulating in Alaska's water sources, as well as radon gas. Other ponds may emit neither, highlighting the importance of identifying not only where greenhouse gases are likely to be released, but new sources of hazards for human health.

Even in attempting to tally these changes, researchers may underestimate nature's complexity. Liljedahl explained that when ice-rich tundra degrades, it can slump and become a pond. As it fills with moss, a very effective insulator, the underlying permafrost sometimes recovers, eventually filling up the depression with a bonus layer of new organic soil. "Instead of losing, it's gaining," she said. "We can't lock ourselves into the idea that it can only go in one direction."

Emerging back into the light, Skidmore stared out over the hills, where pockets of birch marked where mining operations disturbed the permafrost a century ago, creating pools and altering the forest. The catastrophic flooding revealed within his tunnel will happen again, he mused. "It's only a matter of time."

American Motherhood

FROM *The Atlantic*

I LOOKED AT the clock glowing on the nightstand in my bedroom and it read 1:23, *one-two-three*, a neat set of numbers. I tossed and turned and writhed and looked again, and it read 1:17. Had I misread the clock? Maybe I was dreaming about the time. Maybe I was just confused.

I slept, I woke up, I "slept," I "woke up." I hobbled into the bathroom, feeling shooting pain each time I moved my left side. The veins in the stone on the vanity writhed and breathed. Everything smelled metallic. I was hallucinating. I itched, and so I scratched, clawing at the damp back of my knees, my soft belly, my ribs. I broke open the scabs on my legs, watching my blood bead on my irritated skin. Back in the bedroom, a strange pair of eyes, slate-blue with yellow sclera, stared at me in the mirror. I had given birth to my second child a week before, and nothing made sense.

The hallucinations that arrived post-delivery were far from my worst symptoms. I experienced debilitating nerve pain during the second pregnancy—like having a tattoo gun alight on my skin, over and over. I itched unceasingly and uncontrollably during both: For 136 days the first time and 167 days the second, I was itchy every single moment of every single day. The sensation ranged between the tight skin of a sunburn and the agony of poison ivy. The itching intensified after sundown, causing sleeplessness and exhaustion. I was itchy in my dreams. I sometimes wonder whether my son, sharing my body, might have been itchy too.

For me, pregnancy was "obscene," in the phrasing of one of my doctors. And mysterious. Over the course of my two pregnancies, more than forty physicians and midwives, by my count, failed to explain why my blood work kept coming back with so many anomalies, why so many debilitating complications kept piling up in an otherwise healthy woman.

Though my experience was unusual, I did have something in common with countless other pregnant people: Despite recent medical advances, bearing a child remains startlingly dangerous, a fact that America's lawmakers on the bench have chosen to ignore. One in five pregnant people experiences a significant complication. And one in four thousand dies during pregnancy, in childbirth, or shortly after delivering, including one in eighteen hundred Black mothers. Yet Justice Samuel Alito's opinion in *Dobbs v. Jackson Women's Health Organization* takes the interests of the "unborn human being" into account while dismissing those of the person forced to carry a pregnancy to term.

Gone are *Roe v. Wade* and *Planned Parenthood v. Casey*, which protected the constitutional right to termination. Eight states and counting have banned abortions, with no or minimal exemptions for medical reasons. But there is no standard for what preserving the "health" or the "life" of the mother means.

"What actually counts as the life of the mother?" asks the physician Jennifer Jury McIntosh, a spokesperson for the Society of Maternal-Fetal Medicine, an organization for obstetricians who handle high-risk pregnancies. "Is that her life today? Is that her life during this pregnancy? How close to losing her life does she have to be for us to decide to terminate? That feels really weighty, because my interpretation of what I feel is lifesaving—does that align with a particular prosecutor's interpretation?" She told me that she feared situations in which not ending a pregnancy might constitute medical malpractice, but doing so might open her up to criminal prosecution. (She practices in Wisconsin, where nearly all abortions are now banned under a law passed in 1849.)

Worst-of-the-worst stories are beginning to surface in the press, an appalling countdown clock ticktocking until Americans learn of the first woman who dies after being denied a termination since *Roe* was overturned. Media attention focuses rightly on those worst-of-the-worst cases, in which abortion is obviously lifesaving. In countless other cases, the circumstances will be murkier and stranger, but the elimination of the option of abortion will nevertheless do grievous harm.

As it would for me. My two pregnancies left me disabled, a word I am still struggling to come to terms with. They put my life at significant risk. Some of my doctors have made clear that they do not think I should bear a child again. Still, if I got pregnant, I would likely be forced to carry to term in much of the country, despite how sick I was, despite all the damage and pain I endured.

The 2018 midterms were looming, and my husband and I were planning to relocate from Washington, DC, to California, when I learned that I was pregnant the first time. At our first ultrasound, we sat besotted, looking at the gummy-bear contour on the screen as a *wub-wub-wub* sound filled the room. The first jump scare came minutes later. The technician went silent, moving the wand over my pelvis again and again. In time, a doctor came in to tell me there was a mass the size of a softball—roughly the size my baby's head would be at forty weeks, she noted—sitting on the left side of my pelvis. She was not worried about it affecting the pregnancy. But she was worried about it twisting and cutting off the blood supply to one of my ovaries. That would be "a nine or ten out of ten on the pain scale," the doctor said, as flatly as if she were letting me know the time.

Duly noted, I guess. She expected the baby to arrive at the end of March or early April, and with that deadline set, my husband and I hastened our move across the country. Toward the end of my first trimester, as we were packing boxes and giving away books, the itching showed up, as I noted in my diary at the time. I first noticed it in the morning. Lots of people stretch and then scratch themselves when they wake up, right? Next, I noticed it in the evening, a reaction to our laundry detergent, perhaps? Soon there was no noticing necessary. I could not *not* notice it. Especially in warm weather and at night, I felt centipedes scuttling over my feet and hot needles poking my shoulder blades.

Things I tried to make it stop: hypoallergenic detergents and soaps; pine-tar soap; eczema wash; calamine lotion and capsaicin lotion; oatmeal baths; shea butter and dozens of other emollients; antihistamine cream; a variety of over-the-counter and prescription oral antihistamines; topical steroids; showering with cold water; holding bags of frozen corn against my body; wearing loose clothing; acupuncture (not fun when you are itchy); screaming. Only three things helped: ice, a cream called Sarna that made me smell like an unsmoked Newport, and scratching. So I iced, applied lotion, and scratched. I tore the webbing between my toes and the thin skin around my belly button. I clawed out clumps of hair. I fantasized about being able to scratch my wet bones or the jelly in my eyeballs.

My West Coast providers were sympathetic, if not overly concerned. Itching was not unusual during gestation, my new obstetrician noted. She suspected I had intrahepatic cholestasis of pregnancy, a dangerous condition that affects roughly one in a thousand pregnancies. She ordered weekly blood tests to look for a telltale increase in bile acids. In the meantime, she told me to use cold compresses.

Aside from the pruritus—the medical term for itching—I felt joyous as my belly swelled and we got used to our new home, hiking on the beach with our dogs and trying out baby names. My husband and I took the cult low-intervention birthing class everybody we knew took, called, ahem, "Yes to Birth!" I read book after book about how I was built for this.

As the weeks dragged on, though, the itching kept getting worse, and the blood work showed nothing. By the middle of my second trimester, struggling to function, I called my providers, sobbing and begging for help. The nurse on call asked me if I had taken antihistamines. Of course I had. "If you felt like you had poison ivy for months," I snapped, "did you think you would have taken a fucking Benadryl?" She had me leave an escalatory message on some voice mail. Nobody called me back. So we moved on to our third set of medical providers, a team of crunchy, evidence-based midwives in Oakland.

At our first meeting, my new lead midwife took one look at my skin—which, from all my scratching, had undergone a process called lichenification, becoming thick and leathery and covered in welts—and referred me to a dermatologist, who guessed I had either scabies or cholestasis and prescribed a strong topical steroid. (Alas, I did not have scabies.) The new drug would work, I promised myself. I lurked at home, talking to my baby, who we had learned was a boy, telling him I could not wait to meet him.

That happened sooner than I'd imagined. Early in my third trimester, my midwife kept muttering "Fuck" as she took my blood pressure. It had spiked to 170 over 90, a level suggesting that I had preeclampsia. She called in an obstetrician who deals with high-risk cases to take over my care, as I had "risked out" of her scope of practice.

That new doctor, No. 4, confirmed that I had preeclampsia, with anomalies in my liver and kidney labs. She admitted me to the hospital at thirty-one weeks so she could monitor my organ function and asked me to stay there until I gave birth, which was likely to be at least a month early. "We'll deliver you at thirty-six weeks," she told me. "But you're not going to make it that long."

I raged. How dare she! *I* would deliver—not *us*—when my body and my baby were ready. I was healthy. I was built for this. (I was in denial.) Her face softened. She had a feeling about some pregnancies, she told me. Mine was one of them. Then she told me about two patients who had had severe pruritus during pregnancy. The first gave herself frostbite during a blizzard, ripping her clothes off and letting the snow numb her. The second threw herself out a window.

Doctor No. 4 agreed to discharge me if I would come back for monitoring in the hospital every third day, so we did that, and I would cry listening to my son's heartbeat in the dark of the sonogram room. Two long weeks later, I went in and got a 220 over 100 blood-pressure reading, then 180 over 110. Then time sped up: The chatty sonogram technician going silent—*that's not good.* Itching. Machines beeping, cuffs squeezing, panicked phone calls. Itching. A meeting with a gray-faced neonatologist, the decision to induce. IV poles. Itching. It was Valentine's Day, I realized— red hearts, blood-soaked linens. A dozen people rushing into the room. No time to induce, time to move to surgery. "Sign this form." "Stay still." "Your husband has to wait outside." Cutting, tugging, pulling on my insides, as if I were a prey animal being eaten alive. The discovery that my placenta was shearing away from my uterus, a complication that could have killed me and the baby.

I am not sure exactly when I became a mother. He was quiet at birth, and the neonatal team whisked him away before I could see him. We met hours later, my tiny valentine in his glowing plastic box, covered in wires, fed by a tube.

I was eating soft pretzels with my siblings in a mall food court a year later when my younger sister tried to discourage me from having another child. "Don't do it," she told me. "You could have died. The baby could have died. Don't do it."

We did not die, though. My three-pound newborn had grown into a thriving toddler, all wacky smiles and fistfuls of spaghetti. As for me, at four days post-delivery, I was so itchy that I demanded a surgeon amputate my legs. ("We'll take that under advisement.") The next day, the itch disappeared, leaving my doctors and me never really knowing what it was. Months of intensive trauma therapy, started when my son was in the NICU, had restored my mental health in a way I did not think possible at first.

In the end, it was my kid, in all his chaotic wonder, who convinced me to try again. Wasn't it worth it? What wouldn't I sacrifice for him? I prayed that the complications would not come back and that the baby would be safe in my body until we reached full term. But if they did, I knew I had the right to a termination for medical reasons.

When I got pregnant the second time, my providers considered me high-risk from the start. A team of clinicians scrutinized my body and blood, anticipating problems rather than reacting to them. The irregularities started to show up in the first trimester, again—problems with my liver labs, then itching, then nerve pain and insomnia.

My providers could do little to manage my symptoms. As the pregnancy wore on, the aperture of my life drew smaller and smaller. I struggled to think. My hair fell out. I developed gestational diabetes, snatching away the routine of a sandwich, the delight of a good banana. Injecting insulin into the tight drum of my abdomen made me itch every time I did it; I took to stabbing my thighs with my finger-stick lancet after I was done, because pain inhibits itching. I became so foggy-headed that I drove the family car into a concrete pylon. From time to time, I felt so overwhelmed with love that I could not imagine not being pregnant. From time to time, I felt a yen not so much to die as to not exist at all.

Early on, my high-risk-pregnancy obstetrician had called in a hepatologist, a genial older guy with a cravat and an indeterminate European accent. He initially told me I did have intrahepatic cholestasis, both in my first pregnancy and in this one. But well into my second trimester, he left me a voice mail: "I would like to discuss your disease." My *disease*? He had added some tests to my blood work, he explained. My body was teeming with an antibody indicating that my immune system was attacking my bile ducts. I had a disease called primary biliary cholangitis—slow-moving, rare, degenerative, and incurable. The hormone load of pregnancy was exacerbating its symptoms.

Hot astonishment, cold relief. I was sick. I had been sick for years. I would be on medication for the rest of my life. I would likely be diagnosed with other autoimmune conditions in time, and was now more susceptible to developing a host of diseases, including liver cancer. My pregnancies had "unmasked" this reality, as my hepatologist put it. And in creating life for my children, I caught a glimpse of how my own might end: brain-fogged, fatigued, vitamin-deficient, dry-eyed, and, yes, itchy.

But such symptoms might not emerge for years, if at all. After I delivered my son, I hoped, I would again be delivered from the miseries of my own body. The day finally came, and I decided to opt for an induction instead of a C-section, given how much I had hated the latter experience the first time. I labored, feeling the itching subside with the pain of every contraction, breathing myself closer and closer to relief.

Of course, we were in for a few final scares. During one cervical check, my midwife said my son was emerging hand-first. She would have to push him back up to reposition him. (This, well, hurts.) And she worried about his heart rate, which kept dropping, indicating that my contractions might be compressing the umbilical cord and he might have to come out via C-section after all.

A doctor placed an epidural. Beeping, people crowding the room. Pushing. *I thought I was supposed to feel pressure but not pain.* The need to get him out, now. The decision to try forceps. My husband blanching as he stared at the monitor. A request from the doctor for more anesthesia, a warning that it would take a few minutes to kick in. Searing pain, leading me to scream at the doctor to stop. The doctor with the forceps not stopping. Nurses pinning my legs down in the stirrups. The sense of being eviscerated, the room going Technicolor. I did not know I could feel so much pain.

At least it was over quickly. The baby emerged quiet, floppy, purplish. My first reaction was to laugh. I went through all of that to have a baby who was *dead?* But soon he started to breathe, letting out a raspy cry. The extra anesthesia kicked in, so I keeled over in the bed as the doctor explained that he'd performed an episiotomy while wresting the baby out. I was insufficiently anesthetized at the time. "I'm sorry," he said softly.

My experience of pregnancy was over, and I slammed the door

shut. I thanked my clinicians and meant it. I ate three adult portions of spaghetti, ignoring the admonishments of the nurse who had to come in to give me more insulin. I washed away the blood the nurses had missed between my toes and ministered to my stitches and scratch marks. I declined to go back to trauma therapy, though I knew I needed it. I had a beautiful newborn, a beautiful toddler, a beautiful husband. I took my torn-up body home.

Some heartbroken part of me wants to have another kid. I am a middle child, my parents are both middle children, and I am married to a middle child; I have a deep-seated intuition that families are most fun when they are big. Plus, I did heal postpartum. The hallucinations went away when the itching stopped and I started sleeping again. My doctors found a cocktail of medications that have ameliorated many of my symptoms. But I know I cannot give birth again—not least because I need to be a parent to the children I already have and love. And I cannot imagine being forced to.

A majority of Supreme Court justices and a number of state legislators not only can imagine that scenario, but would make it a reality. They would let states force childbirth on victims of rape, incest, and intimate-partner violence; on people without the means or desire to raise a child; even on literal children. The decision is about life, supposedly—the anticipated lives of the embryos and fetuses people will be required to gestate against their will, like broodmares. The lives of those pregnant people do not matter. My experience does not matter. What matter are the opinions of judges and politicians, most of them men with no medical expertise and no experience of being pregnant, no visceral understanding of how messy and hard it can be, even in the best of circumstances.

Even those best of circumstances can be obscenely dangerous, and *Dobbs* has made them more so. The ruling will "absolutely lead to an increase in maternal morbidity and mortality," McIntosh told me. Physicians are now waiting for women with prematurely ruptured membranes to decompensate before providing an abortion. They are reconsidering how to care for pregnant people with cancer, among other ailments. There are reports of people being denied drugs that could be used as abortifacients, even if those drugs are being prescribed for a different purpose.

The judiciary is forcing pregnant people and their doctors to justify care that was once allowed by constitutional right. What if

I had had to plead my own case? Would I have been denied a termination if I had never gotten my proper diagnosis, and had only a history of extreme symptoms and abnormal test results? Would I have qualified on the basis of my liver disease and diabetes, both now diagnosed as permanent? How about my chronic fatigue, or my history of pregnancy-related mental-health troubles? Would a physician have been able to end my pregnancy if I'd had another placental abruption or another bout of dangerously high blood pressure? Would the state have required me to itch for months on end as my body and my mind deteriorated?

These queries are personal, but *Dobbs* has raised more metaphysical ones too. What does the life and health of mothers mean? How could it possibly mean so little? What are we supposed to do?

Contributors' Notes

JESSICA CAMILLE AGUIRRE is a writer whose work focuses on climate change and extremes. Her reporting has appeared in *The New York Times Magazine, Vanity Fair, n+1, Harper's Magazine, The New York Review of Books,* and many other outlets. In 2023, she was a fellow in the Ted Scripps Environmental Journalism program at the University of Colorado, Boulder.

EMILY BENSON is a senior editor for the magazine *High Country News.* She writes and edits stories on science, the environment, climate change, and other topics from her home in North Idaho.

DOUGLAS FOX (www.douglasfox.org) is a freelance journalist who writes about the earth, biology, and polar sciences. His recent trip to Thwaites Ice Shelf in 2019–20 was his sixth visit to Antarctica. Doug has written for *Scientific American, National Geographic, The Atlantic, The Virginia Quarterly Review, bioGraphic, Science News,* and other publications. He is a contributing author to *The Science Writers' Handbook.*

SARAH GILMAN is an independent writer and illustrator based in Washington state. She serves as a contributing editor for *Hakai* and *bioGraphic* magazines, and a writing mentor for the *Open Notebook.* Her written and illustrated work has appeared in *Yes!* magazine, *The Atlantic, High Country News, Audubon, Adventure Journal Quarterly,* and several books. She is also a proud writer-drawer for *The Last Word on Nothing.*

VANESSA GREGORY has written for *Harper's Magazine, The New York Times Magazine,* and *Orion,* among other places. She also teaches journalism at the University of Mississippi in Oxford, where she lives with her husband and daughter.

SABRINA IMBLER is a staff writer at Defector, a worker-owned sports and culture site, where they write about creatures. They are the author of the collection *How Far the Light Reaches* and the chapbook *Dyke (geology)*.

FERRIS JABR is a contributing writer for *The New York Times Magazine* and *Scientific American*. He has also written for *The Atlantic*, *Harper's Magazine*, *National Geographic*, *The New Yorker*, *Wired*, *Outside*, and *The Los Angeles Review of Books*, among other publications. His work has received the support of fellowships from UC Berkeley and the MIT Knight Science Journalism Program, as well as a Whiting Foundation Creative Nonfiction Grant. He lives in Portland, Oregon.

MAGGIE KOERTH is an award-winning journalist and editor who loves to tell stories at the intersection of science, politics, and society. She is the editorial lead for Carbonplan, a nonprofit dedicated to promoting scientific research and accountability to improve carbon capture and storage. Previously the senior science reporter for FiveThirtyEight, she is a Nieman Fellow (2015) and a member of the board of the Council for the Advancement of Science Writing. Maggie lives in Minneapolis with her daughters and her elder cat.

ANNIE LOWREY is a staff writer at *The Atlantic*. She is the author of *Give People Money* and a forthcoming book called *The Time Tax*.

J. B. MACKINNON is a writer as well as an adjunct professor of journalism at the University of British Columbia. He lives in Vancouver, Canada.

BEN MAUK lives in Berlin. He writes for *The New York Times Magazine*, *The New Yorker*, *Harper's Magazine*, *The London Review of Books*, and *The Virginia Quarterly Review*, among other publications.

JOSH MCCOLOUGH's writing has appeared in *The Missouri Review* online, *Epiphany*, *Puerto del Sol*, *Splash!*, *Split Lip* magazine, and *New World Writing*. Josh received his MFA from the University of Iowa's nonfiction writing program. He lives with his family (and Gus the Bernedoodle) in the suburbs of Chicago.

MARYN MCKENNA is a senior writer at *Wired*, a senior fellow at Emory University's Center for the Study of Human Health, and the author of *Big Chicken*, *Superbug*, and *Beating Back the Devil*. She has been published in *The New York Times Magazine*, *The New Republic*, *The Atlantic*, *The Guardian*, *Scientific American*, *Smithsonian*, and numerous other publications. She won the AAAS-Kavli Gold Award for Magazine Writing in 2019.

LOIS PARSHLEY is a freelance journalist and photographer. She was previously an editor at *Popular Science* and *Foreign Policy*. Her reporting on science and geopolitics is wide-ranging and has been published in *The New Yorker*, the *New York Times*, *The Atlantic*, the *Washington Post*, *Harper's Magazine*, *National Geographic*, *Businessweek*, and *Wired*, among others.

MARION RENAULT is a freelance science and health journalist based in Grenoble, France. Their work—which often focuses on extinction, emergency medicine, ecology, and caregiving—has appeared in *The Atlantic*, the *New York Times*, *The New Republic*, *STAT*, *Slate*, *The New Yorker*, *Wired*, and more.

FLETCHER REVELEY is a freelance writer based in Arizona. He has reported from the Middle East, Latin America, and Europe on a range of topics including human displacement, migration, climate change, and public health. Follow him on Twitter @FletcherReveley.

ELIZABETH SVOBODA is a California-based science writer and essayist. She is the author of *What Makes a Hero? The Surprising Science of Selflessness*, and her work appears in *Scientific American*, the *Boston Globe*, *Atlas Obscura*, *Discover*, *Greater Good* magazine, and the *New York Times*. She is at work on a book about the art and science of life pacing.

ISOBEL WHITCOMB is a freelance journalist based in Portland, Oregon, whose work covering lakes and rivers, conservation ecology, and environmental justice has appeared in *Sierra* magazine, *Atmos*, *Scientific American*, and more. They're interested in the way humans and wildlife find new ways of inhabiting the new world we've created.

NATALIE WOLCHOVER is a senior editor at *Quanta Magazine* covering the physical sciences. Her writing has won several awards, including the 2022 Pulitzer Prize in Explanatory Reporting, and has been featured in *The Best American Magazine Writing*, *The Best American Science and Nature Writing*, and *The Best Writing on Mathematics*. Wolchover has roots in London, England, and the Texas Hill Country and now lives in New York City with her wife and daughter. She is currently working on a book about the search for the unified theory of nature.

SARAH ZHANG is a staff writer at *The Atlantic*.

Other Notable Science and Nature Writing of 2022

Elizabeth Kolbert
Testing the Waters. *The New Yorker,* April 11, 2022.

Kea Krause
What Seed-Saving Can Teach Us About the End of the World. *Orion + FERN,* November 16, 2022.

Hari Kunzru
Hard Wired. *The Yale Review,* June 1, 2022.

Krista Langlois
The Giving Trees. *Sierra,* June 12, 2022

Sarah Laskow
America's Lost Crops Rewrite the History of Farming. *The Atlantic,* October 1, 2022.

Andrew S. Lewis
Angry Birds. *The New York Times Magazine,* June 26, 2022.

Jyoti Madhusoodanan
Discrimination Is Breaking People's Hearts. *Scientific American,* June 1, 2022.

Adam Mann
Controversy Continues Over Whether Hot Water Freezes Faster Than Cold. *Quanta Magazine,* June 29, 2022.

Laura Mauldin
Care Tactics. *The Baffler,* July 25, 2022.

Rebecca Mead
Norwegian Wood. *The New Yorker,* April 18, 2022.

Sara Michas-Martin
Black Boxes. *New England Review,* April 1, 2022.

Elizabeth Miller
The Saguaro Solution. bioGraphic, August 18, 2022.

Peter Wayne Moe
Bones, Bones: How to Articulate a Whale. Longreads, February 22, 2022.

Robert O'Harrow
The Message of the Mayfly. *Washington Post Magazine,* September 25, 2022.

B. "Toastie" Oaster
Pacific Lamprey's Ancient Agreement with Tribes Is the Future of Conservation. *High Country News,* October 1, 2022.

Abby Ohlheiser
What Happens When You Donate Your Body to Science. *MIT Technology Review,* October 12, 2022.

Stephen Ornes
Will Transformers Take Over Artificial Intelligence? *Quanta Magazine,* March 1, 2022.

Alexis Pauline
Satellites Stalk Narwhals from Space, But They Still Keep Secrets. *Sierra,* June 4, 2022.

Matthew Ponsford
Why Didn't the Toad Cross the Road? *MIT Technology Review,* July 1, 2022.

Summer Praetorius
The Great Forgetting. *Nautilus,* December 19, 2022.

Chanda Prescod-Weinstein
Becoming Martian. *The Baffler,* January 1, 2022.

Leath Tonino
  American Outback. *New England Review,* December 15, 2022.
Christie Wilcox
  The Web of Life. *The Scientist,* July 5, 2022.
Stephen Witt
  The Future of Everything. *The New Yorker,* December 12, 2022.
Tana Wojczuk
  Fallout. *Orion,* December 7, 2022.

Daniel Wolff
  Why A Marsh. *Places Journal,* May 17, 2022.
Malia Wollan
  An Environmentalist with a Gun: Inside Steven Rinella's Hunting Empire. *The New York Times Magazine,* February 6, 2022.
Charlie Wood
  How the Physics of Nothing Underlies Everything. *Quanta Magazine,* August 9, 2022.

EXPLORE THE REST
OF THE SERIES!

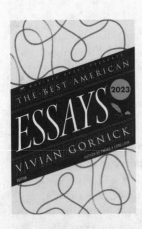

On sale 10/17/23
$18.99